Cover:

Berndnaut Smilde, *Nimbus D'Aspremont*, 2012 -
Courtesy of the Artist and Ronchini Gallery, London, UK

Translation:

Vanessa Place.

She was the first poet to perform as part of the Whitney Biennial;
a content advisory was posted. Her exhibition work has appeared at MAK Center/Schindler House; Denver Museum of Contemporary Art; the Boulder Museum of Contemporary Art; The Power Plant, Toronto; the Broad Museum, East Lansing; and Cage 83 Gallery, New York. Selected recent performance venues include Museum of Modern Art, New York; Museum of Contemporary Art, Los Angeles; Detroit Museum of Contemporary Art; Andre Bely Center, St. Petersburg, Russia; Kunstverein, Cologne; and Whitechapel Gallery, London.
Recent books include *Exposé des Faits*; *Boycott; Statement of Facts*; *Notes on Conceptualisms*, co-authored with Robert Fitterman, her translation of *Guantanamo*, by French poet Frank Smith, and *Last Words,* an artist audio/book.
Place also works as a critic and criminal defense attorney.

With a special thanks to:

Teresa Carmody

1 CITÉ RIVERIN
75010 PARIS
http://www.disvoir.com

ISBN 978-2-914563-80-2

PRINTED IN EUROPE

with the support of

ÉCOLE NATIONALE SUPÉRIEURE DES ARTS DÉCORATIFS, PARIS
- PSL RESEARCH UNIVERSITY

see our complete catalogue on: http://www.disvoir.com

series edited by

DANIELE RIVIERE

in the same series

THE MUTANT FLESH
FABRICATION OF A POSTHUMAN

Denis Baron

NARRATIVITY: HOW VISUAL ARTS, CINEMA AND LITTERATURE ARE TELLING THE WORLD TODAY

René Audet, Claude Romano, Laurence Dreyfus, Carl Therrien, Hugues Marchal

BLACK IS A COLOR: A HISTORY OF AFRICAN AMERICAN ART

Elvan Zabunyan

KEEP THIS SEX OUT OF MY SIGHT

Elvan Zabunyan, Chrystel Besse, Arlette Fontan, Françoise Gaillard, Marie-Joseph Bertini

CONTEMPORARY PRACTICES: ART AS EXPERIENCE

Paul Ardenne, Pascal Beausse, Laurent Goumarre

TECHNO: AN ARTISTIC AND POLITICAL LABORATORY OF THE PRÉSENT

Michel Gaillot, Jean-Luc Nancy, Michel Maffesoli

SOUND AND THE VISUEL ARTS

Jean-Yves Bosseur

IN FAVOUR IN TODAY'S ART

François Dagognet

THE IMAGE-MATTER

EMERGING MATERIALS & IMAGINARY METAMORPHOSIS

DOMINIQUE PEYSSON

Translated from French by

VANESSA PLACE

CONTENTS

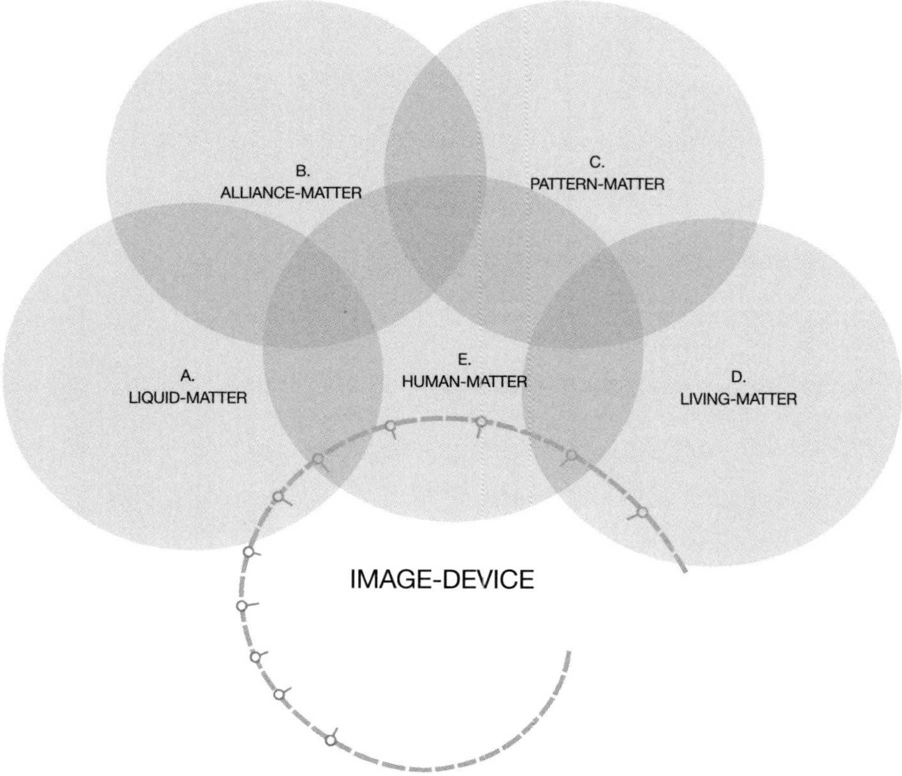

A. LIQUID-MATTER: 1. **Drop-matter** (p.20) - 2. **Micro-liquid** (p.30)

Certain materials have ability to strongly repel water (hydrophobic) or strongly attract it (hydrophilic). These properties are very useful, notably for the creation of drops of fluid isolated inside another fluid. At very small scales, about the thickness of a hair; controlling the liquids allows (for example) the manufacture of drugs, each drop containing a chemical micro-reaction..

B. ALLIANCE-MATTER: 3. **Supra-alliance** (p.41) - 4. **Memory-matte**r (p.68)

Some compound (mixtures of different elements) have been developed to exhibit exceptional properties. The superconductors can levitate objects or carry an electric current with no energy loss. Certain alloys, known as shape memory, can be used to animate the material under control.

C. PATTERN-MATTER: 5. **Infra-structure** (p.73) – 6. **Animated matter** (p.80)

Controlling the internal structure of the material at very small size scales may be used for the invention of new properties. The fine structure of meta-materials enables them to divert light waves to make objects invisible, or seismic waves to protect buildings. Material structured in hyperfine networks can also become ultra-light for exceptional strength or absorption capacities. Their double layered structures allow movement in response to stimulation (electrical, humidity).

D. LIVING-MATTER: 7. **Vegetative matter** (p.85) - 8. **Animal matter** (p.90)

Living matter can be used as material, such as fungi (enlever le deuxième fungi) that grow and take a form directly usable as an apparatus. Such material may also inspire researchers through biomimicry, just as plants and animals have optimized specific functions in particularly clever ways.

E. HUMAN-MATTER. 9. **Invasive invasion** (p.97) - 10. **Trans-mutation** (p.100)

Much research is devoted to making biocompatible material, designed to replace, supplement, or optimize certain functions of the body and brain. Hybridizations between materials introduced into the human body and our own cells are becoming more frequent. The boundaries between the inert and the living thus become increasingly difficult to establish.

AMBULATORY PREAMBLE*

"Let my reading stop and images will now be born of the empty imaginary that had accompanied my reading. However, it is still necessary that this consciousness, which otherwise remains at the edge of the image, find a material that allows it to be realized." Jean-Paul Sartre.

In a fascinating article[1], the philosopher Didier Ottaviani describes the problem of embryology in the Middle Ages. The question of the mystery of life was: how is matter capable of creating its own changes in form? At that time, two distinct approaches coexisted as explanations. The first was that of the artisan and his work — form being external and superior to matter, since it comes from God. The second was that of an internal rapport — a self-generation of form *in* matter. The development of the embryo then became a model system for thought, suggesting the visualization of a new model of the relation of matter/form. Conceived by some as allowing gestational development from the vegetative to the sensory, finally "receiving" the rational soul in late pregnancy, embryonic growth was also interpreted as the combined action of materials with *formative faculties*[2] activated by *energizing powers*. For Ottaviani, the crucial question about the matter-form duality underlying these

* The ideas developed in this writing primarily come from one of three parts of my thesis: "What touches us we touch. Emerging materials are the test of contemporary art. New forms of encountering the sensibilities between man and material."

[1] OTTAVIANI Didier, "Le paradigme de l'embryon à la fin du Moyen Âge", in Astérion [on line]. 01 juin 2003, [accessed February 21, 2016]. Available online : http://asterion.revues.org/25..

[2] An expression used by Didier Ottaviani in his article, on the subject of the embryo according to Albert le Grand.

models was not limited to the illustration of a doctrine. It would also provide a powerful conceptual tool, embryology becoming the vector of a new way of thinking in the Latin world of the late thirteenth century[3].

Obviously we are not here to discuss embryos or the details of this captivating history. But we may be allowed to enjoy this small diversion as a beautiful metaphor that corresponds so well with what we will discuss here. Our relationship to the material is always emblematic of our relationship to the world and all philosophers have sought to try and determine what belongs to the material, what belongs to the sensible world, to the mental, the ideational, the soul ; to establish — or not — a hierarchy between matter and form.

Recent discoveries in fundamental physics do not help us on these points. Subdivided again and again, each particle always thought to be elementary found in others smaller still, definitely causing us to lose the thread of its story. Existing here and there at the same time, mixed with a vacuum that consumes all space, we can never be familiar with the particles that make up matter. If the smallest dimensions remain unthinkable, the mass itself (which essentially defines that which makes material) can become the property of a particle : the Englert-Higgs boson[4].

We have now come as far as describing matter solely as a matrix to be controlled by technical movement, becoming noteworthy only once *informed*. In contemporary art, moreover, the haecceity[5] or "thisness" of material is considered by the artist as both the

[3] *Ibid.*
[4] Detected for the first time in 2012, and explains why some fundamental particles have mass.
[5] The totality of the characteristics which give a thing its particularity and distinguishes it from all other things.

inspiration for and the constituent singular identity of the work. However, the Platonic conception remains valid, the notion of a neutralized malleable material subject to remaking surfacing at a time when technologies of rapid prototyping[6] materially retranscribe the *idea* of a digitally designed form.

The science of materials does not appear revolutionary at first – who doesn't know this – we are used to hearing about crystals, plastics, or strength and or force and resistance in building our homes. Yet laboratory work on the building of matter could well be the source of a new techno-scientific revolution.

These new materials, the emerging materials considered here, are neither amorphous nor homogeneous. They carry their own organization — an *internal form* — which gives them their identity. This internal structure obeys the laws of this material field, not those of God but those of the great fundamental forces. An inner morphology generating exceptional and wholly adjustable properties. These complex materials thus become capable of behaviour (vegetative ?), sensitive to their environment and able to respond (by dynamic forces?). They are capable of inducing their own changes. Some scientists even describe the most sophisticated of them as *smart* —, although we prefer the term responsive. To extend the generative analogy a little further — we will stop there — material controlled by scientists gradually changes status, each day getting a little closer to the living matter with which it is already crossed. A materialistic march towards creation in the ontological sense, that of life and its supplemental soul, perhaps akin to the development of the embryo.

[6] 3D printing, CNC processing (Computer Numerical Control).

The means of observing the material — the primal material, so familiar to us — has not stopped being perfected since the development of the first electronic microscopes during the second half of the twentieth century[7]. These same technological means enable control of structure at such small dimensions that the systems obtained have extraordinary properties. New knowledge, combined with an increased power in modes of fabrication of all possible designs at all scale sizes, provides access to potentialities that our predecessors could not have even imagined. The material thus produced can be transformed into an invisibility cloak[8], artificial skin[9], protocells[10], etc. The high-tech materials become more efficient, become more complex... They mimic biological material and its sophisticated structures and play with the boundaries between the living and the artificial.

It becomes difficult to get a mental image of these modified materialities and their impact on our society, on our relationships with others, on our own being in the world. All these transformations open up new avenues for other potentialities to rush in, breaking the invariants, reshaping the space for thought and confronting the emptiness of our representations. The images we have available to mentally grasp these movements and metamorphoses have become ineffective and need to be redrawn, to be set in relation, to be otherwise arranged. To not let us exceed, invade, smother, or mislead, we need a structure that wears thought, an arrangement of image-

[7] The invention of a high-resolution electronic microscope would not have been possible until the invention of the stigmatizer by Hillier in 1946.
[8] GUENNEAU Sébastien, McPHEDRAN Ross et al., et al., "Méta-matériaux et invisibilité" (partie 1), *Photoniques*, Cahier technique, n°63, 2013, p.50-54. Available online http://www.photoniques.com.
[9] See Organovo website, a company that prints human skin, available on organovo.com
[10] See the TED talk of Martin Hanczyc, The line between life and not-life, may 2011, https://www.ted.com/talks/martin_hanczyc_the_line_between_life_and_not_life.

supports, along the lines of the image-copy and image-fiction discussed by Paul Ricœur. For, as he says, "The main question is no longer how to reproduce mentally (or physically) things in their absence, but how to produce meaning.[11]" Meanwhile, Gilbert Durand[12] introduced the necessary dimension of the sensitive: images are emotional moulds capable of bearing our ideas: they are their predecessors and not their followers. While strange materializations are being invented in laboratories, we will have to cultivate an imaginary able to include them.

For our representations must help us project to the world a clarifying matrix. Even beyond this, we must materialize these mental images because, as stated by Jean-Paul Sartre, consciousness aims at its object through affective analogy: "Consciousness becomes imaging by representing intuitively its object from that knowledge and the emotional material which then becomes imaging knowledge and the material of the image.[13] "

It is in these places that contemporary art can come to play its role as intercessor between matter-form and human thought, between the meta-reality of the work and our social imaginary. Many contemporary works of art produced with high-tech materials have the unique ability to image with great poetry, to carry a strong symbolic charge, stimulating the animated forces that address our perception or confuse our senses. The work of the artist here is to allow *live-ness* to happen: that part of affect which we inevitably link

[11]RICŒUR Paul, *Imagination et métaphore*, communication à la Société Française de Psychopathologie de l'Expression, Lille, 23-24 mai 1981, published by the Comité éditorial du Fonds Ricœur. Available online ligne http://www.fondsricoeur.fr/.

[12] DURAND Gilbert, *Les Structures anthropologiques de l'imaginaire, Paris : PUF, 1960.*

[13] CABESTAN Philippe, *L'Imaginaire*. Sartre. Ellipses. Paris, 1999, p. 17

to the things around us. Also to open the field to conceive-see differently, to leave reassuring but unidirectional tracks. This is the life of our thinking, which will be discussed here. To wander in those places where thinking and weighing walk hand-in-hand[14].

[14] The association of weighing and thinking here refers to the writings of Jean-Luc Nancy, which questions via the intercession of these two words the opposition of soul and body in his book, Le poids d'une pensée, l'approche, Strasbourg : La Phocide, 2008.

PART I — HIGHLY TECHNOLOGICAL MATERIAL

1. MOVEMENT-MATTER

Small, light sculptures are arranged, each in its place, in the exhibition area rising before our eyes. It is a small, simple paper assemblage whose surface arrangement might remind one of plants, all leaves open, or petite insects with great wings, surprised at rest after some fluttering flight. Nothing moves yet: the air is dry. It is when water vapour wets the air that the small sculptures come to life. Each deploys movement in its way, a small dynamic lightness that makes them unfold, move slowly, sometimes startle. Then alight. The air must dry out now to reanimate the little paper show, set in reverse. And then to rest. The alternation of moisture and drought sets the rhythm of the dance sessions, the frail troupe vibrating and acting accordingly, sometimes in one direction, sometimes in another. Some take the opportunity to move forward a step at a time. The ballet is presented according to the same slow beat, each participant playing its part. All moving so gracefully but slightly jerky, like a flower opening or a pine cone quickly shedding its scales. This isn't surprising: the physical principle of the movement of these small papers is modelled on that of plants...

Why describe this small visual artistic experience? Simply, to have it open the dance of our thought concerning the materiality of high technology. Because, surprisingly, these little pieces of paper that respond to humidity in the air are a typical example of the emerging materials that we will discuss here. Of course, it's not just a matter of saying that the paper is an emerging material ; what causes the movement of these small structures is directly related to ingenious arrangements of the material at multiple scales of size. If the paper surfaces come into motion, it's because they consist of two sheets of different materials: one paper, the other of another material. It is their apposition, providing what is called a "bilayer," which constitutes the mechanics of this small dynamic. Because each layer does not react in the same way in a humid atmosphere. One swells with water, as much as possible. The other ignores it, refusing any change in volume. But their common surface leads them into a direct conflict whose resolution requires some contortions. Indeed, it is the inflection of their surface that will allow the two sheets to correspond while responding to their strong internal injunction, causing them to inflate or not. It is the correctly chosen connection between these two types of surfaces that allows the structures to actuate according to certain rules. The choice of structure translates into a type of movement. And the passage of moisture to dryness, or its reverse, provide the power for this engine.

This example shows the importance of the internal structure of the material, in all size scales. The typology of motion is provided by the choice of assembly of the material layers. But the internal structure of the paper material also intervenes in smaller dimensions on this swelling: it does not occur in the same way in the direction of the alignment of fibres during manufacturing or in an orthogonal direction. The paper therefore does not react in the same way along its principal axes... Similarly, closer examination of the fibres reveals a complex arrangement of cells that are not at all homogeneous. The swelling of these cells causes expansion of the material. Things do not stop there, however, and researchers have shown that inside the cell, walls can be found which have the same for arrangement of molecules, conducive or unconducive to swelling. We will not go into the physicochemical details of this experience, but will remember this : the chemical composition of the material is not the only thing responsible for its reaction. The structure of the material, in any size scale, is crucial in obtaining its sophisticated properties.

But extend this reasoning to all of today's high-tech materials. It is the specific arrangement within the material, made possible by its control to nanoscale dimensions, which allows it to attain properties the natural material does not possess. In 2006, for example, a researcher at Imperial College, London, John Pendry, created a structure with a negative refractive index : a laboratory curiosity that the theorist Victor Veselago predicted in 1967 but which was not possible to make at that time. John Pendry and Ulf Leonhardt offer this in the creation of an invisibility cloak from their new metamaterial: offer this potentiality water bypassing a stone in a stream, light avoids these structured materials rather that being obstructed. Researchers are also developing a super-lens whose resolution is no longer limited by the classical laws of optics. Research has been focusing on these incredible discoveries and some laboratories have actually been able to realize a cloak of invisibility in the microwave regime. Many researchers, sponsored mostly by the military, which is very interested in camouflage, are working to control manufacturing in the infrared and visible spectra. The difficulty lies in controlling the structure at the scale of the molecules (ten nanometers), as the material needs to be composed of both metallic periodic structures formed of concentric[15] rings and continuous metal threads. The envisioned solution for manufacturing a periodic system at such small dimensions is that of directed auto-assembly : a technique that is still in the experimental stage. The metamaterials thus obtained do not exist in nature[16]. Their action on light and on the world around us, exceeds — by far — all we have experienced to date, thanks to the fine structure which constitutes it.

Consider again our little laminated papers, and two of their peculiarities.

The first is that they obey an injunction – of a somewhat unusual sort – which is that water is conducted to them through air. Not unexpectedly, but as conditions of humidity, an environment to which they respond. It is therefore possible to control them, at least in part, by wetting or drying the surrounding atmosphere. We will call theses materials responsives, as they respond to changes in their environment.

The second feature is their similarity to plants. Scientists speak here of biomimicry, inspired by the ingenious solutions which nature finds to achieve its ends. Contiguous blade systems benefit from this *movement without muscle*[17] that plants have invented for their use. The principle is plant based, but can be specifically reinterpreted for other systems : those of small, lightweight papers, large wooden plates, or other systems yet to be invented: this area of research is beginning to be developed. The architect Achim Menges uses this process for his responsive constructions : openings in the sides of his wooden building *HygroSkin* (2013) are clovered by thin

[15] They are called split-ring resonators. (SRR).
[16] This is a material which has, for example, a negative refract ve index, a medium of infinite impedance or relative permittivity of less than a unit.
[17] MWM : Moving Without Muscles.

strips of wood. They open like flowers when it is warm and close when it rains. A sublime exploration by this artist-architect of other ways of imagining the habitat. Our artistic investigation of bilayer systems has been done with the help of physicists Philippe Marmottant[18] and Etienne Reyssat[19], who are concerned with the generation of objects that have a wide deployment of movement modalities. These systems develop their own behaviours, depending on the form of their movements; we then like to create typologies that, through projection, lead us to see the viewers as beings endowed with life.

2. EMERGE

Developments in the field of materials are so spectacular that some practitioners, carried away by their enthusiasm and positivism, predict a change of civilization at least, a saving transhumanity at most. Without going in these extremes, it's safe to say that this century will be largely about making materials[20] of a new kind. The changes these materials will induce, from the more imperceptible to the more resounding, will be enormous. Our relationship to *material things* and perhaps especially to the *living* — or even to "living things" — will necessarily have to evolve. But what material is it ?

The name may already identify the contours of the field covered by these new materialities. We choose to call them *emerging materials*. As defined by their membership in a new phase of scientific advancement in more closely modulating the properties of matter — seemed both sufficiently precise and encompassing. In the last

[18] Researcher at the laboratory of the Interdisciplinaire de Physique de Grenoble.
[19] Reaearcher at the laboratory of Physique et Mécanique des Matériaux Hétérogènes à l'ESPCI, Paris.
[20] "The twenty-first century could be called the 'Multifunctional Materials Age'": excerpt from the back cover of the book, Smart structures theory, CHOPRA Inderjit, SIROHI Jayant, New-York : Cambridge University Press, 2013, 896 p. (Livre 35).

a-

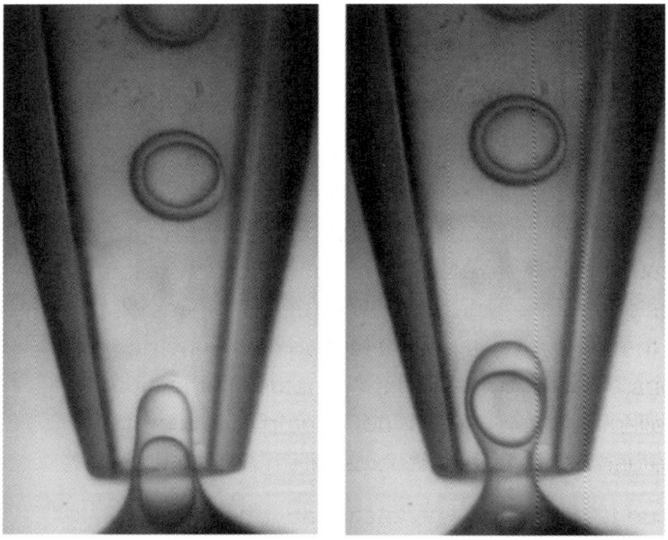

1. Drop-Matter

a- Inclusion of drops rendered immiscible thanks to molecules with hydrophilic or hydrophobic properties. Researcher and photographer Laura Adams.

b- Water drops on a hydrophobic surface. Grains de Bâtisseurs (video excerpt) www.grainsdebatisseurs.com/

c- Super hydrophobic and oil repellent material used as a coating to repel liquids and keep the objects clean.

d- *Hydrophily*, fogging installation. Dominique Peysson, Hsinli Wang, 2013. The drawings are obtained by working the surface of the support, in order to encourage or discourage fogging. With the Scientific council of Daniel Beysens, EnsadLab supports..

e *Hydrophily*, detail.

b-

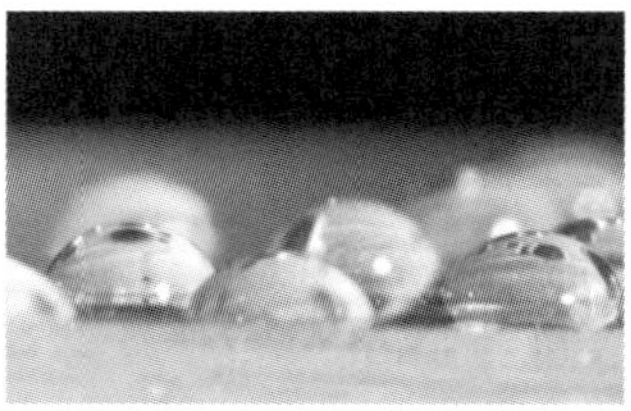

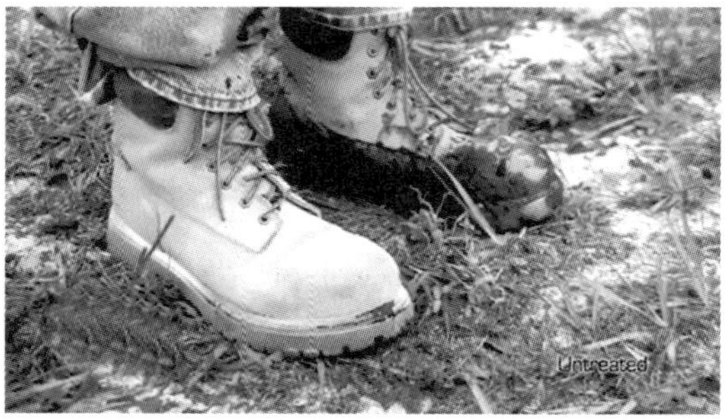

c-

d-

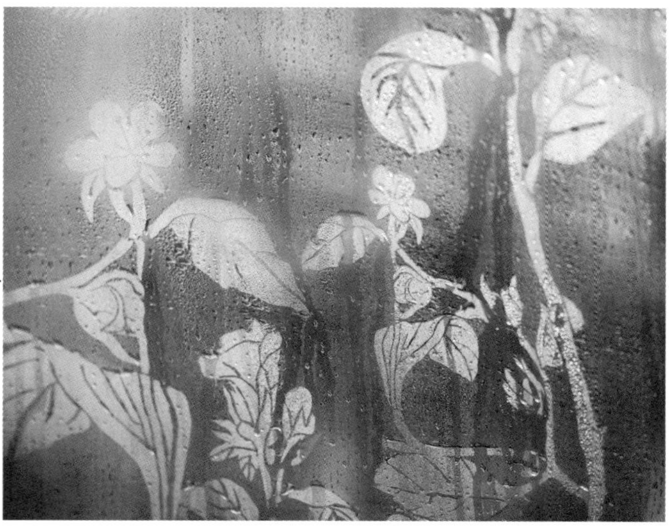

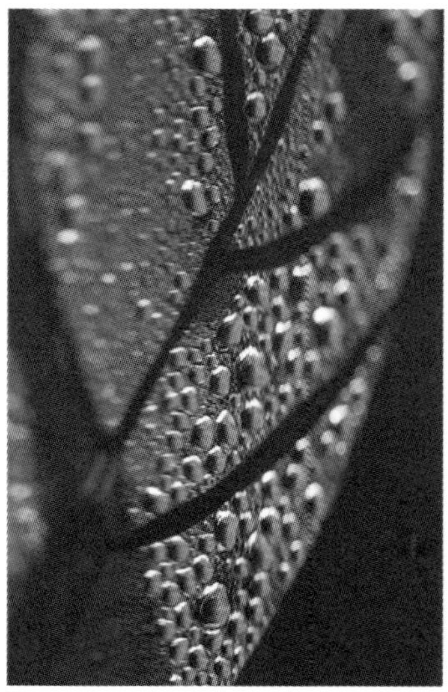

e-

century, science has completely transformed our way of seeing 'things' and developed an entirely new approach to the material. The term 'emerging' reflects our criteria: an *emerging* material fresh from the laboratories and proudly bearing its complex material condition — this emergence being (and we will come back to this) what characterizes most clearly the principle of physical complexity. Of course, it is not their novelty (or rebirth) that interests us, although it is fascinating to be faced with a proliferation of new materials in their profusion of progeny and metamorphic mutations. It is their special, optimized performance that we will consider here; the properties of emerging material *emerging* above those of all others. Their flexibility, elasticity, lightness, thinness, transparency, colour, flexibility, delicacy, strength and, of course, biodegradability.

Included within this *emerging material*, according to our definition, are all materials with a particular and optimized property which at minimum differs distinctively from those of natural materials (stone, wood ...), or conventional materials: metals, ceramics or glass, and new resins and plastics. Their characteristics can occur as much from their chemical composition (elements, molecules) and their fittings as from their large or very small scale structure. These materials are often complex and high-tech.

Among these *emerging materials*, we will give pride of place to that fascinating class that some call smart — "*intelligent materials*", in French — or less excessively, "*adaptives*". The English term is distinguished from the adjective adaptive in that it introduces anthropomorphic qualities such as *sensitivity*, *openness*, and *enthusiasm*. These materials are both sensitive to their environment and are capable of responding. A material that perceives and reacts, but whose abilities end there; ending with intelligence: we prefer to call them *responsives.* This word is not used in French, although it is mentioned in the dictionary as an adjective for *answer*. The English version (responsive) is commonly used in architecture and design to characterize objects or their design that detect certain environmental properties and transform accordingly. A *responsive material*, according to our definition, thus a material sensitive to its environment (light, temperature, pressure, humidity, etc.) which responds in remarkable fashion.

Some of these innovative materials can now, thanks to their particular functionalities, converse with the living. These developed materials become increasingly complex, each day coming a little closer to that of the living. The inert sometimes takes its place and the border between the living and the artificial grows increasingly tenuous.

3.PATTERNS AND INTERLACINGS

Emerging materials express their capacities in many different registers and the possibilities available to us have multiplied, becoming almost endless. These new materials turn to gold in the hands of creators from all backgrounds, provided perhaps — for the non-scientist — that they take the trouble to tame them in the laboratory. Using these materials requires a preliminary step which is essential in making contact with them. We could say to *think with them,* or even, perhaps, *like them*... This is done by some contemporary artists, whose appropriations of certain of these fabulous materials give us a glimpse of the power of some material-thought, this thought-image, this image-matter. This requires feeding our imagination with the founding principles of materials science and their imaginings.

Austere and sublime

The matter is a system consisting of a large number of unit-elements, organized (or cleverly disorganized) in all size scales, clusters, molecules, atoms, particles, etc. Materials science consists of establishing relationships between that which is made and that which is making with the properties that it can be made manifest. The organization of these units and the relationships they establish between themselves is thus fundamental when characterizing their differing types of materials.

Crystalline matter, for example, differs from other, less regular matter, by its perfect ordering. The strictly identical repetition in crystal of the same pattern (atoms, ions or small molecules) arranged according to a geometry exactly reproduced throughout the material, leads to the extremely regular structuring of a lattice. The set of properties can be determined from knowledge of the smallest characteristic unit, its unit cell which, repeated endlessly in the three directions of space, generates the crystal. Here we face order in its pure state, as one might speak of pure spirit. The materials regarded as ‘purest’ are moreover mostly crystalline in nature: precious stones, cut so as to isolate a single crystal, metals — such as bronze composed of copper atoms and crystallized tin, stones and ceramic materials, also used for thousands of years. Granite, meanwhile, owes its name to the particular appearance of the multitude of small crystalline grains of quartz, feldspar, and mica.

The systems called organic in everyday language are those assigned the symbolic power of life: a set *out of control*, which is positioned as the antithesis of the abstract and its purity or hardness. No straight lines, no points: all direction is embodied in a volume whose convolutions mingle and transform indiscriminately. But the classifications of scientists do not exactly cover this imaginary of the living. In chemistry, the organic can be said to be any substance composed of atoms usually present in living organisms, such as carbon, oxygen, hydrogen, and nitrogen. Small organic molecules can, for example, crystallize and present a perfect periodic structure.

Rather what counts lies in the assemblage of material elements. This organization is to be understood as the order (or disorder) internal in the material and its structure. The philosopher Jean-Philippe Narboux emphasizes this fine distinction between order and structure as two complementary but not superimposed notions: "The organization determines the identity of the system, whereas structure determines how its parts are physically articulated.[21]" Jorge Wagensberg, meanwhile, provides a particularly vivid and expressive example of the importance of structure in the organic. Referring to a scene from his childhood, he says: "[...] the water molecules, it's clear, are not the same as those of old, nor are those of the plants nor those of the dragonflies, nor those that cause our olfactory or tactile sensations. Nor those of our fingers! Thirty years ago, there were other particles, atoms, and molecules. Since then, the material has been replaced thousand times. What persists then with such a strong evocative power? This is not about the particles but the relations between them, their arrangement...[22]". The

[21] NARBOUX Jean-Philippe, "Diagramme, dimensions et synopsis", in *Penser par le diagramme. De Gilles Deleuze à Gilles Châtelet*. Vincennes: Presses Universitaires de Vincennes, 2005, p. 159..

[22] WAGENSBERG Jorge, *L'âme de la méduse. Idées sur la complexité du monde*. Paris: Le Seuil, 1997, p.10.

organic can be made in convolutions, in cascading folds, and in turbulence, but is nonetheless hyper-structured, if only through the arrangements of atoms linked by chemical bonds. The degree of abstraction lies in their organization: the immanence of pure geometrics, at once austere and sublime. Or, conversely, their attraction to the clutter so representative of nature as it surrounds us.

Sweet disarray

The simple crystalline/organic duality must be surmounted in order to reach a more nuanced view. The researcher Pierre-Gilles de Gennes, Nobel Prize winner in Physics and great innovator, has invented such a new field of exploration, called soft matter by physicist Madeleine Veyssié[23]. Such soft matter acts as intermediate materials between fluids and solids. They are capable of responding very intensely to low stimuli from their environment due to their ability to rearrange through cooperative effect. Many emerging materials are part of this category. This is particularly true of systems consisting of very long molecules – macromolecules, or polymers – such as many biomolecules. They mingle and tangle with others, unable to align together smoothly as do crystals. And thus, remaining entangled even in their solid state. Interlaced. They constitute a large part of living matter, or the organic, such as wood, leather, wool, and plant fibres. Their particular structure was detected at the very beginning of the nineteenth century, but only in the early twentieth century did we understand their chemical synthesis. They are now present everywhere: the plastics that we all know are macromolecules. The concept of interlacing can also be used to grasp the nature of another range of materials that are solid but not hard : elastic materials such as

[23] See the chapter about soft matter included in the book written and edited by Édouard BRÉZIN and Sébastien BALIBAR, *Demain la physique*, Paris : Odile Jacob, 2004.

natural rubber. If only few molecular bridges lightly attach strings relative to each other, preventing fluency, these elastic materials will never flow like a liquid, regardless of the temperature to which they are heated.

A characteristic of these macromolecules is that these creators can also serve as a matrix. Small molecules with selected properties may be fixed by a chemical bond to the length of these large chains, or just scattered throughout its mesh. The compound thereby obtained will have the plastic properties of the matrix and be able to take advantage of the unique functional properties of the grafted molecules.

Some contemporary artists address the concept of matter not as a solid, yet malleable substance that can withstand time but as a direct expression of fragility, evanescence, or inconstancy. The material is not necessarily a solid and stable body but fluid and flowing material, a vapour which can fill space or condense itself, a passage of matter in various states. But in all these cases, the disorder is total. The mixed molecules roll over each other in the liquid, held together only by forces which prevent them from becoming steam. If the temperature increases, however, they will agitate enough to counterbalance their desire for cohesion and fly freely in the air. They are cut off, they have taken their freedom, to be re-connected by other links or interactions. The material then prescinded from the complexity it knows in its densest state.

Of course, nature has reserved the pleasure of playing at the interfaces of these different states of matter. Thus, there is a liquid state which refuses to give up a certain ordering: the liquid crystal. In the manner of a game of Mikado (pick-up sticks), whose rods align against each other so as to spread more easily in space, the liquid crystal molecules, in elongated or disc form, are preferentially oriented one to the others, although they may move as any other liquid. Another material with amazing properties consists of a fluid plus particles of a sufficiently small size so that the system remains stable whatever the conditions. The combination of the fluid nature of this material, combined with the magnetic property of particles comparable to tiny magnets, produces a new state of matter : the ferrofluid. If we approach this fluid with a magnet, it will deform, even bristling with very regular spikes if the field is strong enough.

4. THE NECESSITY OF COMPLEXITY

Like the universe itself, with its history of continual resistence to the grand physical forces that pull uncontrollably towards general homogenization, the material must be composed of parts that communicate and react. And that these interactions affect the internal structure able to transform it. Life indeed is maintained only through this same necessity : to differentiate. Cannot every living thing be ultimately described as a skillful system which strives towards increasing the treasures of complexity, as sophisticated embedded systems that challenge, grow, destroy, exchange, multiply, interrelate, oppose, then transmit unto death, wherein everything is then newly dissolved into uniformity? Interactions that target one objective – which communicate – to preserve complexity and to sustain life. Communicate to differentiate, differentiate to live.

The great change in the way researchers apprehend matter lies in consideration of its composition and its internal structure. For it is from the nature of this interior "form" that material properties will follow. The arrangement and structure of the constituents of the material, of their different parts together, their diverse and dependent relationships in varying proportions : all these parameters must be considered in understanding the effectiveness of these materials. Each combination producing the specific and unique properties of the material concerned.

But the effects of those relationships are not easy to determine as the level of complexity reached in these systems is very high. Each element acts on many others, which then act in reverse back to the first, a system of entangled relations leading to the unpredictability of their effects. A difficult problem, which scientists could only find effective solutions to after an important epistemological rupture and a transformed worldview. The hard sciences and the humanities have had to integrate new concepts in order to understand differently those systems that implicate an infinite number of crossing interrelationships, and design a new science : that of complexity.

The concept of complexity was introduced in physics to describe multi-component systems, set in inseparable relations. It is the active network of relationships between elements, at once organized and organizing, which generates a defined set. At the end of the nineteenth century, Henri Poincaré, studying the trajectories of the sun, earth, and moon, showed that the most absolute determinism could lead to unpredictability[24]. The problem had been

[24] JACQUARD Albert, "Albert Jacquard (1/5) – Hommage, Interview with Jacques Lacarrière ", in *À voix nue*. France Culture, 24 September 2013, 20h, Director: Nicole Salerne. Repeat of the 20/09/1993 program. In collaboration with Rafik Zénine et Claire Poinsignon. 27 mn

considered extremely simple: if only sun and earth are considered, without any further interaction between the two, the trajectory of the earth would be very easy to predict, as an ellipse with the sun as its centre. But add a third body, the moon that revolves around the earth and all the respective gravitational interactions between sun, earth and moon, and it is no longer possible to write an explicit equation for the positions of each in relation to time. The fate of the three bodies is well defined in absolute terms but it is impossible to predict their position in some future time, for then we would have to know their initial positions with an absolute, and therefore inaccessible, precision. Although the interactions are known and explicit, the system evolves in an unpredictable manner. This discovery definitively signalled the end of determinism... and the beginning of our freedom![25]

Great discoveries have definitely shaken the edifice of the structured knowledge of classical physics, touching the subtlest heart of the matter: unpredictability. The *simple* or the *indivisible*, considered the basis of all knowledge, has finally been revealed as composed of an infinite number of other bodies also in relationship...

Accounting for systems' complexity was carried out in three distinct phases during the twentieth century. The first wave took place after the First World War, with the principle of Holism proposed by philosopher and statesman J. C. Smuts. The hypothesis was that natural objects are wholes but that we can neither produce them by a simple assemblage of parts nor "account for their characteristics and behaviour.[26]" This assertion, applied to complex systems, maintains

[25] The system of the three bodies is exciting, because it has a very long (and complex !) history; in the Poincaré incident, the question was : is the solar system stable long term? A history full of scientific upheavals and errors, which seems to stabilize—for an instant—with the work of the researcher Jacques Laskar, who demonstrated in 1989 that it is chaotic (that is to say that we can't determine its state beyond a certain time).

[26] SIMON Herbert A., Les sciences de l'artificiel. Paris : Gallimard, 2004. (Folio Essais)

a-

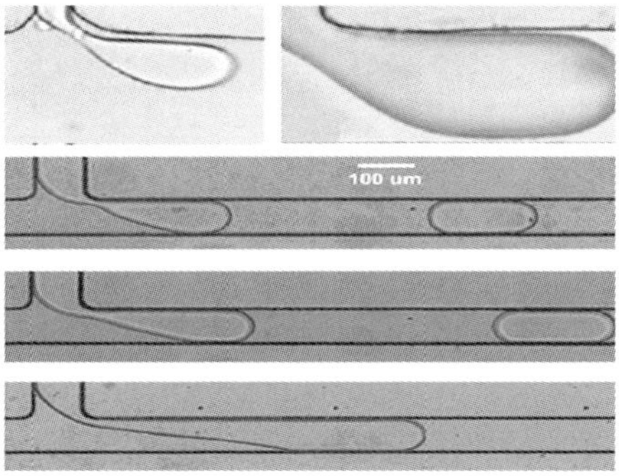

b-

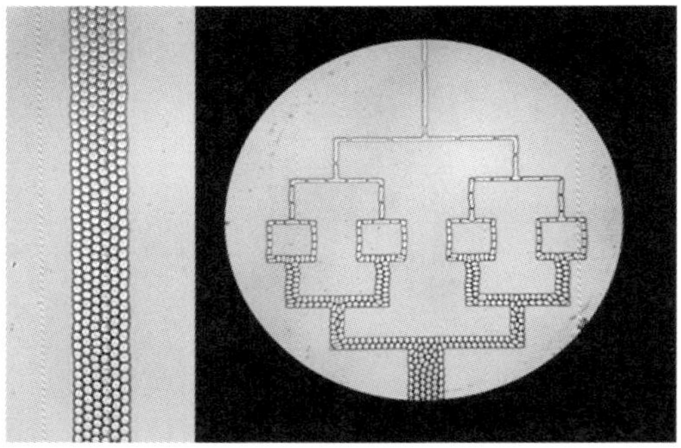

2. Micro-liquid

a- Microfluidic device (microscopic size, about the thickness of a hair). T-junction used to generate droplets included in another fluid. Photo: Brian Carroll and Carlos Hidrovo.

b- Split of drops in a microfluidic circuit, or the art of manipulating fluids on a small scale (10 to 100 microns). Photography David Weitz.

c- VereTropTM, a chip testing laboratory for the detection of 13 tropical infectious diseases using a single drop of blood. Companies A*STAR and Veredus.

d- *Traffic*, microfluidic installation. Dominique Peysson, 2014. *Traffic* project at a big scale moving material in microcircuits as if in a movie. It links the design of the circuits to their operation on the drops, and the passage from analog to digital. Design with Adrien Bonnerot. Scientific Council Patrick Tabeling and Fabrice Monti (ESPCI). With the support of EnsadLab.

c-

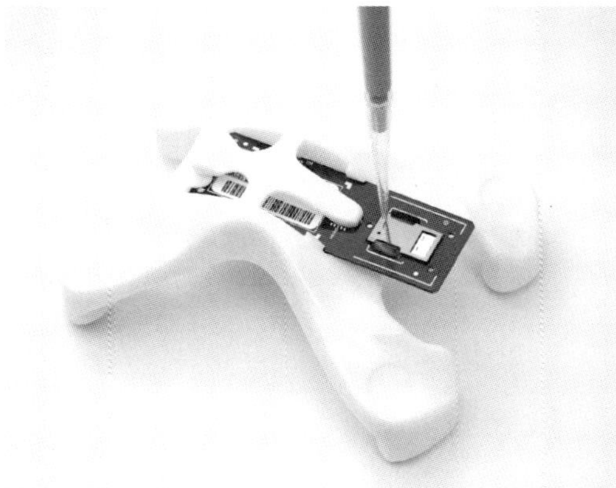

d-

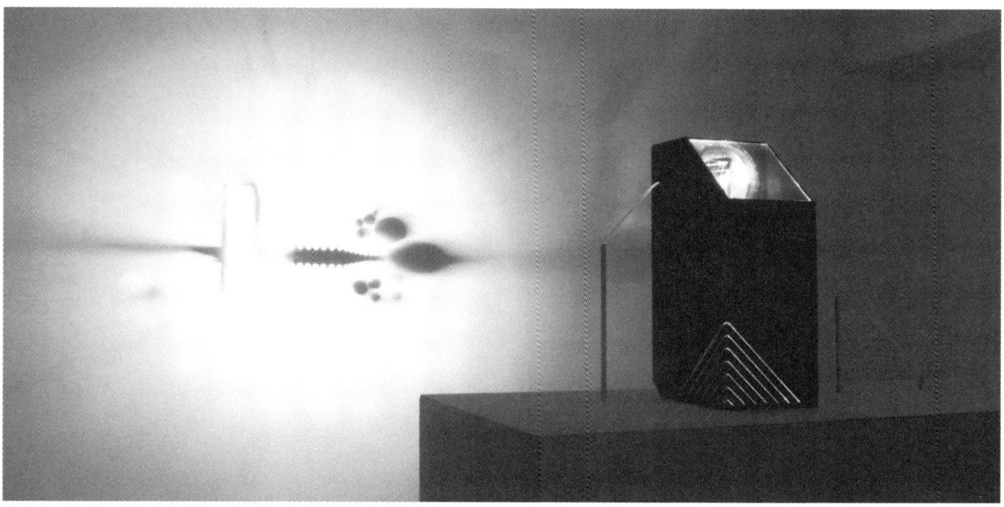

that new properties of a system do not belong to its components but to their interactions and therefore must involve an additional principle, that of emergence[27]. According to *Holism*, emergence is linked to *vitalism*: every living thing is animated thanks to a higher principle, a *vital force* which non-living matter does not possess. Thought, for example, cannot be reduced only to neuronal functioning and calls for an additional stage, which would be this notion of emergence. But for current physics, the emergence is definitely different from that of vital force : there is emergence when a set of parts results in a general behaviour that would not exist if these parts were not united and interacting.

It was at this time that the theory of form, Gestaltism, posed its postulates which reformulated some related concepts in another context: psychic activities take place in an open system, wherein each subsystem is determined by its relationship with its meta-systems. The psyche is the seat of relations between psychological elements that will work towards the emergence of harmony, a perception and conception precise and clear, to compose a good *form*. This is also the era when Henri Bergson published *Creative Evolution* (1907), in which he opposed the principle of the *life force* (*vital impetus*) to the finalist conception of philosophers (and their idea of a metaphysical final cause) and the mechanistic view of modern scientists. Evolution would be neither planned nor predictable but in constant invention.

The second advance in the principle of complexity took place after the Second World War with *cybernetics*, which considers *information*, *retroaction* (*or feedback*) and the *general system*. It was

[27] Emergence is a concept that characterizes a property which follows from more fundamental properties but to which they are not reducible: "the whole is more than the sum of its parts." Emergence was developed in the 19th century by British philosophers (George Henry Lewes and John Stuart Mill in the beginning).

through the initiative of Norbert Wiener that cybernetics developed in the 1950s. Artificial intelligence saw the light of day and Weiner's developments came to contradict the principles of holism and presenting materialism as finally sustainable.

Norbert Wiener transformed social modes of representation, including those our scientific community. Taking into account systems with a large number of components, cybernetics considers interactions between different units, rather than their natures, to determine the system's overall behaviour. Systems, self-regulating, can be modelled only by studying all the information flowing between their parts. While one might think these complex systems as diverse as a social network and a material composed of several molecules have nothing to do with each other, they share common characteristics, in particular the transfer of information, heterogeneity, and the structure of a *small world*. These structural and relational analogies lead to invariant behaviour inherent in complex systems, such as saturation effects, threshold transitions, etc. The non-linearity resulting from feedback loops is a disturbing example. Eluding historical descriptions of physicists who wanted effects to be proportional to their cause (linear effect), feedback loops reflect effects whose magnitude is no longer proportional to their cause. Minute variation of a single parameter can lead to a catastrophic loss of control with disproportionate consequences.

The third stage, which continues today, involves additional concepts that more specifically describe the great systems of the planet. A necessity motivated by the desire to understand problems such as the environment, globalization and living organisms.

Noteworthy were the introduction of *catastrophe theory*[29] (which takes into account and describes the effects of sudden changes), the notion of *chaos* (systems that have the appearance of being random while being well defined and deterministic), *adaptive systems* (able to adapt to their environment through learning), *genetic algorithms* (which permit searching for solutions through natural selection among populations of potential solutions), *cellular automata*[30] (a programming method of dividing space into smaller units and calculating the state of each according to its neighbours, not by time). Complexity becomes particularly high when we enter the domain of the living. Identified as a mosaic, living complexity has the characteristic of assembling complex parts which are themselves complex (cells, organisms...), while allowing each part great functional autonomy[31]. The *Whole* can then be composed of several stages at different scales whose effects combine with the highest degree of complexity imaginable. Biological systems, moreover, introduce the idea of a hologrammatic system. The hologram is an image obtained by a particular optical system which allows points to be defined such that each contains all the quality, relief, colour, and presence of the system as a whole. This is the case with cells, which possess the genetic program of the whole body in their nucleus. It is therefore possible in this to say that the part is not only in the whole but the whole is in the part.

[29] Catastrophe theory was founded by mathematician René Thom for studying dynamic systems that can engender morphology, or discontinuous systems. It enables taking account of and describing the effects of sudden transformations.

[30] SIMON Herbert A., *op. cit.*, p. 306-318.

[31] In art, a mosaic is a representation whose perception does not preclude the appearance, if you look closely, of small elements that retain their shape, colour, and their brilliance. This idea, proposed by Georges Chapouthier, is useful in describing the mechanisms of vision, human evolution, ecology, and embryonic development. See CHAPOUTHIER Georges, “Communication: des neurones aux populations animales”, in *L’Archicube*. Mai 2010, n°8, p. 43.

5. MATERIAL-TECHNOLOGY

An other high-technology

All emerging materials are the result of extensive scientific development involving advanced technical expertise. They are thereby able to fulfil one or more specific functions that enable manipulation of processes which would otherwise be impossible. The combination of the terms in *material-technology* may prove fruitful when faced with the definitions of new technologies found in countless writings on the coming reign of technology. *Material-technology*, however, is perpendicularly defined as high-tech. Unlike the virtual world invoked through an interface or by capture/ response systems governed by "if, then..." laws, the *high technology* of the *material* is an integral part of our tangible world and addresses our senses above all. It is an expression of its uniqueness as *technological material* that increases this material's power to engage our perceptual apparatus.

World-making

The question of the device in contemporary art, after Walter Benjamin, has allowed certain issues to be updated, including the relationship between the *work of the sensory* and the technical *passages* capable of making history[32]. For philosopher Pierre-Damien Huyghe, the "arts are related to these passages, they are their conditions, they prove themselves aesthetically"[33]. Again it appears that *material-technology* (that of emerging materials) is capable of

[32] HUYGUE Pierre-Damien, on the subject of his book *L'art au temps des appareils* [personal website], [accessed on october 7 2014]]. Available at:
http://pierredamienhuyghe.fr/publications/lartautempsdesappareils.html.
[33] *Ibid.*

making social changes to the same degree as the devices studied thus far. Just as the camera and the video camera are now accompanied by a particularly diverse line of multi-media instrumentation, the *new materials* used by artists in the twenty-first century will certainly be the source of new technological families which are poised to multiply exponentially. They are — as efficient materials — new modalities of sensory engagement. As argued by Huygue, these devices are able to demonstrate a strong "power to give the world, to make the world"[34], providing the means for the artist to recreate in the context of the exhibition another reality, that of the artist's interiority. Unlike digital instruments, however, which art has appropriated little by little, emerging materials are rarely used by artists. Not yet widespread, they are, perhaps, still the side of the sacred.

The material as "technical object"?

Technosciences, so called by Bruno Latour, thereby define a fresh look at science in the making, and designate the joint work between theoretical and symbolic activity and technological production. Scientific research is not limited to this practice of representational contemplation, but is involved in the construction of reality. It has been 50 years since philosophers like Gilbert Simondon and Pierre Ducassé deplored society's inability to symbolize techno-scientific operability. The articulation of being-in-the-world and of contemporary techno-scientific civilization was inappropriate, and it became necessary to develop a *technological culture* able to symbolize contemporary techno-scientific developments. Today, this analysis seems obvious, but the difficulty of its realization has only

[34] *Ibid.*

continued to increase, making practical implementation particularly demanding. These are the future productions of scientific research that are hardest to symbolize as they exceed the framework of our individual and collective imagination.

To learn and understand the phenomenological advent and development of technology, Gilbert Simondon devoted himself to apprehending its materialization in the form of *technical objects*. For him, it is this the description and analysis of their mode of existence that allows definition of their essence. According to Simodon, it is not the practical functionality of these objects, including their uses and usefulness in various circumstances, that should be considered authentic. It is rather their diversity and their variation, as well as how each individuates, that enables their identification. Because we need to be able to adequately capture the variation in their characteristic similarity. It is the idea of genesis, evolutionary nature in the genetic sense of the term, that has proved most suitable. What is special about the technical object is very unstable, so it is more efficient to find the laws of its genesis within the frame of its individuality and specificity[35].

Would it not be wise therefore, to engage with *material-technology* as with *technical objects?* Because complex materials are designed not as a *set of defined materials* but more as a *continuum of possibilities* within each family of materials. In the same way as a sculptor, who sizes his material through successive approximations towards a morphological goal, who pours molten material into a mould that he has crafted, who models his material with his hands in the clay, who gradually builds his piece by assembling its successive

[35] SIMONDON Gilbert, Du mode d'existence des objets techniques, Paris : Aubier-Montaigne, 1969, pp.19-20

elements or who practices digital prototyping and 3D printing, a researcher can now *act* in many ways on material to *sculpt the properties* desired. At all scales, from the macroscopic to the atomic. The mixture (at the smallest scale) of particular material entities and combinations of different materials to produce a composite allowing for invention and modulation of an infinity of distinct materials with extremely diverse and expertly optimized properties. Their applications are not all assigned in advance, and like the technical object, to better understand its technicality, we need to understand its "implementation process"[36].

Observation of the material at the atomic level via various scanning tunnelling microscopes and scanning probe microscopes has radically transformed our knowledge. This knowledge, combined with an increase in the ability of manufacturing methods to control all possible arrangements at all size scales, has led to development of a material's potentiality in a continuous process similar to that in designing conventional technical objects. Extending the length of a molecular chain may, for example, be seen as the counterpart of the elongation of a mechanical piece in the design of a device.

There are also families into which emerging materials fall: quasicrystals, nanomaterials, gradient materials, metamaterials, cellular materials, superconductors, foams, liquid crystals, shape-memory materials, biomaterials, etc. Their number is so large that it is impossible to list them all. But our point may be illustrated through the now well-known example of carbon nanotubes. Belonging to the family of nanomaterials, they are a good example of what we consider as a material of the future. Their development is particularly complex

[36] The first chapter of the book by Gilbert Simondon "Du mode d'existence des objets techniques" in *Genèse des objets techniques: le processus de concrétisation.*

as it involves creating long tubes made of graphite sheets[37] wound around themselves, whose total length can reach a centimetre.

These nanotubes share certain characteristic properties such as a mechanical strength comparable to that of steel while being much lighter. However, their principle of development, can create tubes with very different properties as some can be more conductive than copper, or become semiconductors, depending on their geometry. Some can generate colossal electrical fields at their extremities, able to extract electrons from matter, then emit them in a precise, ultra-directed point.

The various applications lead not only to the development of new materials, but also to new assemblages, thereby introducing an additional degree of complexity. The systems thus designed, around composites, cannot be seen as a small-scale version of our machines, or highly valued products of human ingenuity, because the forces involved are not the same and the possibility of intervening in the material at this scale depends on having the tools to do so. The lines obtained by molecular extension will never be comparable to the lines of a device with an elongated mechanical part, primarily because of the physics that applies to this level of complexity. The forces and interactions that occur within material at very small scales are of a different type than those we usually experience, making it possible to modulate the electronic, magnetic, mechanical, or chemical responses of the material at the source. The dimensional reduction of the units from which new materials are designed may be used to invent properties previously considered unattainable. A nanostructured material such as highly purified copper obtained through electrodeposition can be

[37] Carbon in natural mineral form.

stretched hundreds of times more than conventional pure copper. Conversely, certain properties may result from the carefully controlled distribution of impurities in the material (doping), as in the case of the semiconductor used in microchips. The design of multifunctional materials, therefore involves a different philosophy than that of the mechanistic age, though perhaps returns it to its sources of inspiration, – if one considers the flying machines of Leonardo da Vinci, that is to say: nature. As the machines mimicked the mechanics of bird flight, these new materials are often bio-inspired.

One family of emerging materials is very different from the others: biomaterials. Designed to interact with living systems, they can assist or replace, or diagnose such systems, and are most often optimized in a medical context. They must therefore be bioinert, sometimes capable of actively functioning in the biological system in which they are implanted. In any case, they greatly contribute to dissolving the boundaries delimiting the living from the inert. Being completely artificial or derived from living matter grown in the laboratory and then transplanted into the body or hybridized with the living, justified based on biocompatibility and auto-renewability, biomaterials produce a strange intermingling between the inert, the living, and the artificial. Technological processes, such as 3D bioimpression now permit generation of highly complex tissues capable of replacing skin or the cartilage lost in serious accidents.

In this family it is also possible to consider all scales of size. At the molecular level, manipulating DNA molecules is becoming increasingly easier. In 2012, for example, a mechanism was discovered in bacteria which allows changing the genes of living organisms at will. It has therefore become possible to modify the genome of organisms in an increasingly controlled manner. The ethical merits of this and its applications will have to be carefully framed, such as with the reprogramming of cells, the modification of non-reproductive somatic cells in reproductive cells, or even modifying the DNA of an embryo...

There is another way to understand the principle of *tuning* the material. Researchers at the MIT MediaLab, including Hiroshi Ishii and Jean-Baptiste Labrune, are working. For example, on the physical materialization of digital data[38]. The principle of their work posits a different conceptual level: rather than controlling material in its inner structure, material entities of a given form and size may be designed to be manipulated as a modular being. This type of project, like *Radical Atoms*, engage extremely malleable dynamic physical units, coupled with a digital underlying model, such that change in physical form can result in change of digital signal in real time and vice versa[39]. The intelligence of the material may thus be coupled directly with a numerical formulation according to procedures yet to be refined.

We can infer the path of society from the steps of these researchers. We must develop a biodegradable composite material, minimize its price throughout its life cycle, and increase its performance capacity. Proposed materials must be able to adapt their structural characteristics, to master their lifespans and their capabilities, to continuously diagnose their conditions, to provide for their self-repair, to manage their morphostructures, and to perform controlled movements in diverse operating conditions[40]. This is still a self-reflexive loop, in which performance materials become able to greatly influence our perception of the world and, consequently, the forms given our conception of new materials.

[38] ISHII Hiroshi, LAKATOS Dávid, BONANNI Leonardo, LABRUNE Jean-Baptiste, "Radical Atoms: Beyond Tangible Bits, Toward Transformable Materials", *Interactions*, Vol. XIX.1, 2012, p. 45. Available at: http://tangible.media.mit.edu/.
[39] *Ibid.*
[40] CHOPRA, Inderjit, SIROHI, Jayant, op. cit., p.1-2.

a-

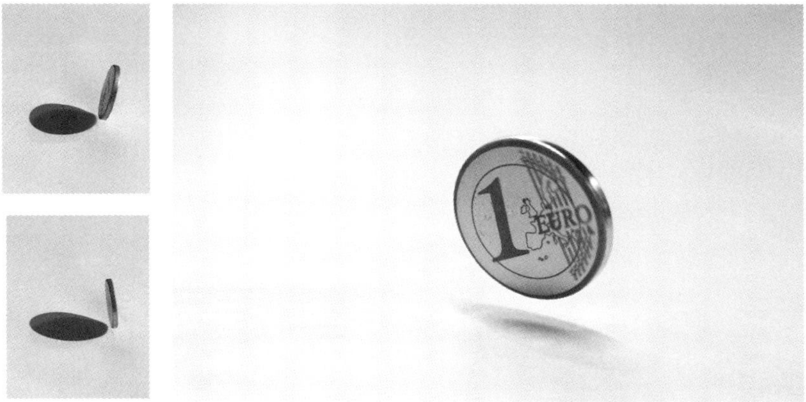

b-

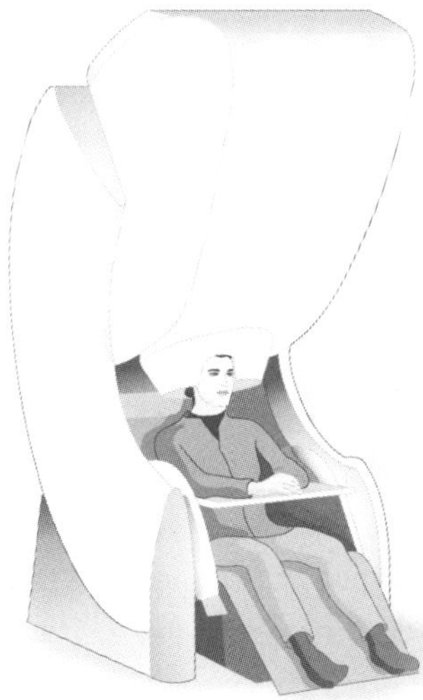

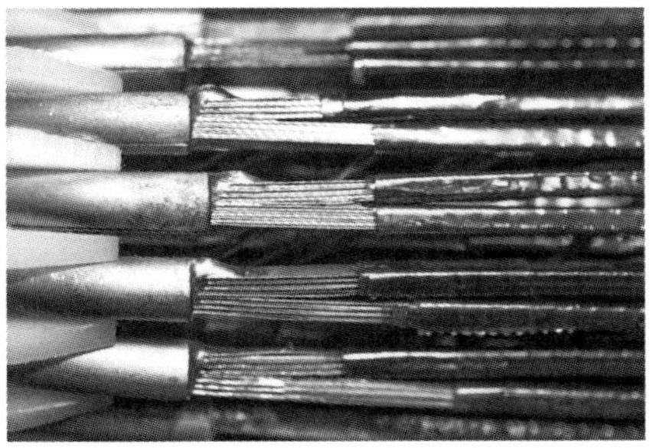

c-

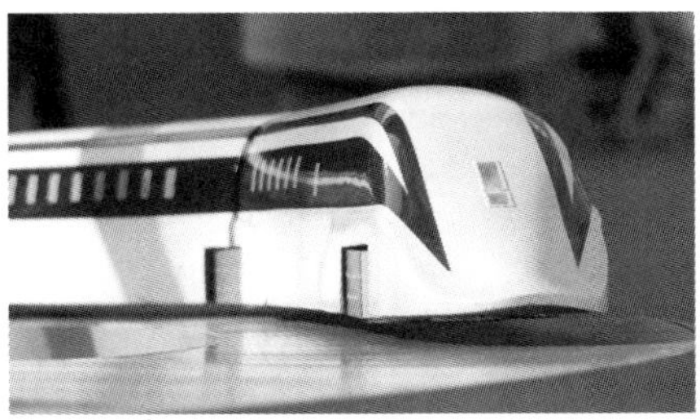

d-

3. Supra-alliance

a- *Or,* Levitating Coin, performance installation. Marie-Julie Bourgeois, 2014. A coin levitates using the superconductive powers of a magnet in a state of hesitation, oscillating between heads and tails before falling. The device illustrates an object subjected to opposing forces that act upon it, keeping it in balance, making it wobble.
b- Magnetoencephalography (MEG) is used to draw maps of the brain with a temporal accuracy of a millionth of a second, using superconducting sensors. This allows the study of problems related to anomalies in brain synchronization.
c- Companies, like Umitomo Electric Industries Ltd, have put on the market high temperature superconductor cables which have the ability to conduct current without loss via the Joule effect, provided they are kept at the temperature of liquid nitrogen (-196 ° C).
d- The MAGLEV is a train levitated and propelled at high speed (about 600km / h) by superconducting coils, kept at very low temperatures. An operating line is scheduled for 2027 between Tokyo and Nagoya in Japan (286km).

PART II — MATTER FOR THOUGHT

"The images found by men evolve slowly, with difficulty, and we understand the profound remark of Jacques Bouquet, 'an image makes as much work for humanity as a new trait for a plant." Gaston Bachelard[41]

1. IMAGINARY MORPHISMS

Hylomorphism: That old couple, material-form

***M**aterial* is a philosophical term that defines itself in opposition to *spiritual*. All we perceive through the senses would be material. More generally, one can say that material is that which has mass and fills space. The term, however, can have a slightly different connotation when used to explain the substance of which something is made, here referring to the concept of *formatting*. This materiality is therefore understood as an extended substance that can be divided into parts and which can be given a form. For a million years, humans have used the elements at their disposal as raw materials: wood, stone, bone, horn, and hide. The Neolithic, marked by profound technological, economic, and social changes, extended this pallet with clay, wool, and vegetable fibres, and then later with the first metals. Only since the first industrial revolution has the scope of what is possible grown exponentially.

[41] BACHELARD Gaston, *L'eau et les rêves*, (1942), 12e éd. Paris : Le livre de poche, 2012, p.9.

Somewhat paradoxically, while the material generally refers to the tangible — commonly with a pejorative connotation, through differentiation with the pure spirit from which it is detached — it can figuratively re-join that spirit against which it was defined in antinomy. The material, which so prosaically visits the slums of our condition as creatures of flesh and bone (ancient medicine spoke of the material to discuss any fluid discharged from body orifices, such as vomiting or faeces)[42] may also rise to the level of spiritual food. While constituting the informative substance from which our cognition and imagination develops, it becomes *material (food) for thought*.

However, in the history of ideas, the material that makes up each parcel of the world will be confronted more or less submissively with the ideational form that defines its contours. The coupling of matter and form — hylomorphism — is one of the oldest operational concepts in our history of ideas. But the emphasis culturally accorded to form, or the material, has long been very unbalanced. It is not helpful to revisit in detail the historical origin of this formalist conception. Idealist philosophies that have profoundly influenced our reading of the world are obvious. If Platonism has demarked the spiritual on this point, it certainly responds to the need to seek control over the potentially maddening upheavals of the sensory world. For Plato, the beautiful comes from good and truth, and the three principles are inseparable. Without this triad, the world would be chaos. Intelligibility is accessible through the ordering of the world,

[42] Contemporary art will will continue to explore this disturbing aspect of material which enters or exits us, a subject totally repressed by our contemporary society :society: « I've always thought of the project as a sort of sexually driven digestive system, that it was a consumer and a producer of matter. And it is desire driven, rather than driven by hunger or anything like that. » (Mattew Barney). (*J'ai toujours conçu ce projet comme une sorte de système digestif sexuel, comme étant à la fois consommateur et producteur de matière. Et il est induit par le désir, plutôt que par la faim ou tout autre chose de ce genre*).

which cannot come from too much diversity, from the melange, from the sensual, the phenomenological. The practice of art and its confrontation with the material, is therefore worthless in itself, art having no interest unless it creates the momentum leading to the unique principle which gives consistency to being, to truth[43]. Aristotle also considered form in a privileged manner: form orders the material and defines the object, its essence and perfection. It is form that gives meaning to the material, and establishes the principle of unity within every being. The Platonic idea of beauty and art was taken over by Hegel in the late eighteenth century. For him, philosophy of the mind is the reconciliation of finite nature and free thought, and art appears as an intermediary that achieves this under a sensible appearance. The work of art will create a path between the material and the immaterial, towards pure thought. Elevated to a state of pure appearance, art detaches from brute reality and sheds: reality, shedding its.

In-formation by information

In our time, we remain very attached to representations based on this dissociation of matter and form, even as their relationships evolve. As we have seen, these are built on complexity theories. The generation of form obeys laws that intersect under informational control. Take for example, the case of embryogenesis or development. How do cells that divide and multiply go from egg to embryo and then to foetus, how do they organize themselves to realize THE form, that of being alive? Or how does a simple cell lost in the primordial ocean evolve over the years to give rise to current living forms? For each

[43] CAUQUELIN Anne, *Les Théories de l'art*. Paris: Presses Universitaires de France, 1998, p. 16. (coll. Que sais-je?.

birth must go through the stage of a single undifferentiated cell: a return to the source before recreating complexity born of simplicity. Life appears as the arrangement of that which is inert, at very small size scales: of molecules or even atoms. Because no specific atom is alive, all exist in inert matter. These are the genes present in each cell, which lie at the origin of differentiation, following a schedule of very precise growth[44]. Not a plan for a final system, but an orchestration in space and time of growth or disappearances to come. The genes do not code each aspect of the final form, as biologists thought in the twentieth century. It is the ensemble of genes which are expressed differently according to the position of the cell in the body. In this, they obey instructions sent by the Hox gene, the gene of development. A particularly marked similarity between all Hox genes of all organisms indicates a strong resemblance between the basic plan of vertebrates and invertebrates, establishing an incredibly high relatedness of all living things. Form and matter are inseparable at this level as they simultaneously intervene at every within the slightest increase in growth. The DNA molecule is both material and origin of the shape. Form and essence of the form are closely related as are, as we shall see, matter and spirit.

Analogue material

The composition of emerging material, its arrangement, its transformation in accordance with the laws that regulate it, these material-forms therefore, can be an excellent support for analogy. If we consider brain function, the step of recognizing analogies is similar to pattern recognition[45]. The psychology of form in the early twentieth

[44] COMBES Claude, *Évolution : les grandes questions*, Paris, Le Pommier (poche), 2010 - p. 18-19.
[45] CHANGEUX Jean-Pierre, *Matière à pensée*, Paris : Odile Jacob, 2008, p.188

century based on the fact that perception and mental representation organize phenomena as forms, that is to say, as structured sets and not as a broad conglomeration of juxtaposed data. In the introduction to his *Essay on the Imagination of Matter, Water and Dreams*, Gaston Bachelard clearly distinguished between "imagination that gives life to the formal cause and imagination that gives life the material cause or, more briefly, the formal imagination and the material imagination."[46] In addition to images of forms, there exist direct images that the eye can recognize and the hand can touch. Here, our material-forms scramble things up a bit, allowing interfacing and mixing for the two imaginaries can come together to cooperate. All emerging matter remains true to the description of Bachelard, leading us gradually to the mystery of the unfathomable. But it is also capable of eliciting a miraculously inexhaustible force, that of growth[47]. Poetic images are born rooted within the material: mental images that remain mere formal games and do not touch us emotionally do not survive for long. Any "philosophical doctrine of the imagination," Bachelard tells us, "must first study the relationship of material causality to formal causality".[48] It is this same bipolarity, this differential, dynamic harmony of opposites, that provides the driving force required for rich and creative thinking.

Emerging materials, able to animate themselves, able to respond to us, may, via installation art, speak very directly to our imagination through the play of multiple analogies. For Bergson, the essence of beauty and art has its source precisely in our capacity for sympathy.[49] A sympathy we feel because we create analogies. "We

[46] BACHELARD Gaston, *op. cit,* p.7.
[47] *Ibid* p. 9
[48] *Ibid* p. 9
[49] BERGSON Henri, *Leçons d'esthétique, Cours II* in *Cours I-IV*, Éd.par H. Hude, Paris, P.U.F.,1990-2000.

place ourselves with other human beings in the presence of a material that appears to have nothing in common with us, neither in thought nor intelligence nor activity. When this material in specific cases seems to express an idea, a feeling, or manifest an effort [...] [things] become more like us, we are interested in them as something human, we sympathize with them, we love them, we find them beautiful."[50] A sympathy that can be seen as "an anthropomorphic illusion similar to that of the child who animates and humanizes virtually everything he sees."[51] The material, although inert in the sense of nonliving, can, through its expressive complexity, stimulate thought composed of both affect and intellection. For Bergson, always, intellectual effort also involves imagination, but imagination through the "dynamic schema". This schema exercises an attraction or impulse on images: a voluntary vertical movement. The instinctive imagination is somehow a walk among mental images, and interpenetration of these images occurs according to laws "that are to the imagination what logic is to thought."[52]

The most obvious analogy that this material conjures is, of course, that of the body. Philosopher François Dagognet (who defined himself as a materialogue) said that the body is a supra-materiality. For "the body is material that is united, where all elements are interdependent."[53] This is an association often made with regard to materials that evoke corporality more directly and instinctively. Thus, touching a flexible material in which it is possible to feel an inner hardness in places (although the overall form is not anthropomorphic)

[50] *Ibid*, p.42.
[51] *Ibid*, p.41.
[52] BERGSON Henri, *Le rire, Essay on the signification of the comic*, free publication online, http://www.ebooksgratuits.com/html/bergson_le_rire.html.
[53] DAGOGNET François et ROBERT Martine, "Entretien avec François Dagognet", *Le Philosophoire* 3/2003 (n° 21) , p. 7-16.

results in a feeling of the uncanny.[54] For the material reproduces the touch of a body, of skin and muscle on the bones, even as our eyes do not recognize it as such. Emerging materials understand this, and associate materialities of different regimes. Matrixes upon which are grafted functional molecules may, therefore, assert operative properties. Some materials are particularly well optimized to play the part of skin; others have dynamic capabilities that recall artificial muscles. Still others are arranged to optimally bind hardness and elasticity under the laws of tensegrity[55], as is the case for the whole of our body. Of course, the material can be used to decouple some of the peculiarities of corporeality, some analogies, like touch, or some mechanical properties, so to isolate them in another form, not illustrative. One way to decouple certain affects related to corporeality to better integrate them as in an art installation.

Order and categories

> The use of logic is necessary for intelligibility; the surpassing of logic is necessary for intelligence. Edgar Morin

It is essential to let the mind adhere to the material support of analogy: because analogies and concepts are the basis of all thought[56]. Historically, philosophers have understood their importance: Plato and Aristotle found analogy constructive, if you were vigilant against being deceived by appearances. Kant, too, believed in its

[54] The uncanny is a Freudian concept, which characterizes the unease (seen either as anxiety or hysteria) that is triggered by an object animated by a movement, without knowing whether the object is alive or not, or by an inanimate object that makes one wonder whether or not it will animate.

[55] Tensegrity is a principle of architecture, but also valid for many other systems like the human body or biological cells, according to which a set of rigid and nonconnected parts are compressed by elastic parts in tension which then links them. The set constitutes a rigid system that absorbs all the demands made on the system over the entire system.

[56] HOFSTADTER Douglas, SANDER Emmanuel, *L'Analogie cœur de la pensée.* Paris: Odile Jacob, 2013, p.9

creative power, and Nietzsche, whose thinking otherwise differed so strongly from Kant, proposed the image of a mobile army of metaphors for that of Truth. Bergson, as we have seen, incorporated the concept of analogy as thought-support. Not analogy versus fixed terms, but in reference to movements or trends: our own inner movements or movements of the universe in general. It is not a matter of similarity between the two terms of the analogy, but a shared identical dynamic.

Categorize for thinking

The notion of order is fundamental for describing different material states. By analogy, we can say most scientific arguments are based on the concept of order. Indeed, the need for organization of our mental material is essential when looking to achieve intelligibility. But of what *order* would be this *order* in which we need to think? "Order," says Michel Foucault, "is both what is given in things as their inner law, the secret network according to which they see themselves, and what exists 'through the gate of a look, attention, of a language'[57]."[58] This is what structures the mental space in which we can think things relative to other things, because of our culture, our life experience, our memory, our perceptual habits, our language, our values... The linguist Ray Jackendoff[59] follows this same thesis when he says that information produced by language is not related to the real world but to the projected world, i.e., the world as it is *organized* unconsciously in our mind.

[57] FOUCAULT Michel, *Les mots et les choses*. Paris: Gallimard, 1966, préface, p. 7-11

[58] Michel Foucault continues : "The fundamental codes of a culture, those governing its language, its exchanges, its techniques, its values, the hierarchy of its practices—fix at the outset for each man the empirical order which he will face and in which he will find himself." . Idem

[59] JACKENDOFF Ray, "De la syntaxe à la pragmatique", *Nouveaux Cahiers de Linguistique Francaise* Geneva : Droz, 2006, n°27. Available at: http://clf.unige.ch.

Hofstadter and Sander explicate this process[60] from the idea of the mental category: "a mental structure that evolves over time, sometimes slowly, sometimes rapidly, which contains information in an organized form that permits its access". Categorization is this operation of associating, in a temporary and gradual manner, an entity or situation to an existing category[61]. These categories, which we effectuate and constantly reassess, are going to depend on our language but also on our culture, our time, and our moods. Each category's definition seems clear and reliable, due to its familiarity and unquestioning daily use, as is the case for each individual word that we use. Though not actually reflecting reality, words create the impression of quickly dealing with situations, rendering them intelligible.

These categorizations rely on analogies, to which Hofstadter and Sander give a broader definition than the one usually assigned[62]. This is a particularly important principle, since analogy, far from being occasionally used, is used continually, incessantly, up to several times per second, at the start of cognition, in order to access intelligibility. Activated from the most basic recognition of a familiar object, it is the first process in the most innovative artistic creations, and the grandest scientific discoveries. Categorization, associated with analogy, permits sensitivity to similarity, similarity, in order to understand novelty and strangeness. A mode of operation that takes advantage of the infinite richness of past experience. Each category, when brought to our attention, inferentially activates[63] a network of

[60] Their work is in the field of psychology and cognitive science.
[61] HOFSTADTER Douglas, SANDER Emmanuel, *op. cit.*, p. 21.
[62] *Ibid.*, p. 22-30.
[63] For Hofstadter and Sander, the inference in psychology is also taken in a broader sense than that usually used in the field of artificial intelligence. Here, inference refers to information that leaves our "mental torpor" to be brought to our attention, to activate our minds.

associated information: wood is a category that opens our mind to many images, ideas and sensations, such as that of a noble material which we like to work, of polished furniture that we love to caress, of trees and a forest in which we like to wander, or the good smell of a fire which we light in our heath.

Categorization “reigns over thinking at all levels, from microscopic to macroscopic[64]”: a conversation is structured from ideas, that govern images, and phrases, which then induce expressions, words, and, finally, even letters. A process in scales, reminding us of the imbrication of size scale in the material. This principle of thinking by analogy allows abstraction to be manipulated particularly effectively by linking an abstract idea to a concrete situation or an object. Commonplace expressions demonstrate this well, such as getting your hands dirty, making ends meet, stamping the feet...

Cognitive flexibility also largely depends on our ability to move from one level of abstraction to another and to reconcile or differentiate situations by effectively determining *what will count*. Scientists frequently use this elsewhere: take, for example, the description of the progression of sound as waves of air pressure. It was developed from observing waves at the surface of water after a stone has been thrown. This theory then enabled Huygens to propose the first theory of undulation of light. Analogies also strongly supported Einstein in building his theory of relativity, where an actual image could be of no use.

The evocative power of artistic works takes the same principle, supported by inferences issuing from objects and the

[54] HOFSTADTER Douglas, SANDER Emmanuel, op. cit., p. 36.

materials they contain. Relationships may be established directly between matter and ideas, as abstract as they are, without going through the usual stages of hierarchical logic. It is mostly at this level that artists' work is positioned. Installations and their components can act according to *analogical levels* operating as contexts for the "coalescence of the sign, the similarity of the signifier and the signified, the homeomorphism of images [...][65]"; a kernel of condensation, said Beuys, speaking of his work using raw materials, "upon which many things can settle.[66]" Condensation that acts as a principle of continuity between different levels of reality by preserving the *real* character of relations between every thing.

Following Beuys' lead, use of emerging materials allows for creation without necessarily separating "essence" from "appearance". The physical process and the substrate material may become truly spiritual principles for working this (abstract) material of thought-perception. A site-source of the development of thought that could be conceptually located before language, still free of any anchor to the symbolic assigned by words. The arbitrary approximation (that of the artist) of different elements constituting a work develop a network of relations by mixing its related inferences. The uncertainty provided by the artist encourages the viewer to create from his own inner being his own system of meanings.

Lotus Dome (2012) is one of the responsive works of Daan Roosegaarde[67], having the particularity interesting feature that it

[55] BARTHES Roland, "Le démon de l'analogie", *Roland Barthes par Roland Barthes*. Paris : Le Seuil, 1995, p. 50.

[56] Excerpt from an interview with Joseph Beuys in 1970 conducted by Jörg SCHELMANN and Bernd KLÜSER, reproduced in: "Questions à Joseph Beuys", *Joseph Beuys: The Multiples*, New york : Schellmann, 1997, p. 19.

[57] More exactly semi-responsive, semi-interactive, according to our criteria, as it directly puts into play an alloy of memory of form, equally used by collectors and actuators.

utilizes smart metal, probably a shape memory alloy: the artist does not reveal his secret. The dome structure is higher than a person and consists of a tessellation of hundreds of flowers made of polished metal whose open petals close the surface. However, light from the centre passes through the interstices and sends out a multitude of finely combed rays. When one comes to *Lotus Dome*, simply putting a hand to the structure will open the nearby lotus petals like a flower and let in the light.

A relationship that isn't tactile, but is almost tactile, epidermal in any case, which plays in the delicacy of the movement of the thin sheets of silver. The response of the material is straightforward: it does not pass through the actuation of a motor. It appears to be initiated by the heat of the hand, and this is what the studio video presentation suggests. It seems more likely that sensors detect the presence of a hand and send a strong, warm light in its direction. This is the temperature rise produced by the lamp which allows the petals to open. Regardless, it is heat from the lamp that opens the petals in a non-mechanistic movement, that of matter. This poetic sculpture invites a meeting of a special kind, since it engages the haptic sense which is rarely used in museum work. The texture of the surface with its openings like skin, the pores opening to reveal the intimacy of its being. An analogy of movement that recalls Henri Bergson. The alliance between the promiscuity of the encounter and the large size of the dome produces a very powerful effect. The exhibition sites of *Lotus Dome*, mostly great and sublime architectures, such as the church of Sainte Madeleine of the XVI in Lille, amplify this symbiosis between the grandiose and the sensitive. The spirit — His light — and the tactile presence, both subsumed into a single point, of which the *centre* is *Lotus Dome*.

Again, the petals' sensitivity to heat allows them to respond when a viewer approaches the work. Temperature represents more than physical warmth: it is a sign of presence. The opening of the leaves associated with this heat is related to spiritual heat. At the source of the sculpture's movement, heat becomes in fact a thermodynamic quantity (to that is to say, linking heat and dynamics). It can flow through the structure, consistent with the movement of the person approaching. Located in the centre of installation and radiating to the furthest walls of vast exhibition spaces, it can be symbolically seen as being at the origin of all life, as source point for the great evolutionary march. It can reactivate primitive links connecting man to the original energy of the earth, from which it takes shape and the animal or vegetal world whose morphology it mimics. Roosegaarde's leaves do not evoke a dynamic, they put it into practice by leveraging a correlated motor principle. The multiple that constitutes the unity, the All, is an essential component of the system : the openings are like cells of a single body, whose behaviour one seeks to capture. Here, the question of complexity is posed in the simplest and most accessible possible terms. Emerging materials, more generally, can be used for their remarkable features but also for their combinations. They allow projection of both the natural thermodynamics of our daily lives and most hypothetical repositories of new operations. They are the expression of this vision of the transformation (or transformability) of materials, of the human spirit and of social existence.

The order of chaos

Our thinking does not consist only of categorization: it builds relationships and connections between what it is able to establish as objects. But they are not always rational links: thought consists of managing multiple interactions between intelligible and unintelligible components that access our understanding. Thought is an informational process that functions in the form of dynamic relations between our logical reasoning and phenomenological inferences[68]. The phenomenological gives meaning to the informational, which would otherwise be a simple calculation. Similarly, phenomenological processes alone cannot give rise to thought: feelings of *being* or of *existing* do not correspond to any structure.

In his book *The hidden order of art*[69], Anton Ehrenzweig explains how a mode of thought can create bridges between feelings and rational thinking. Between the intelligible and the unintelligible. This mode, that of syncretic thought, is the thought processes used by young children before they are able, as they get older, to analyse and rationalize their world. Everything is placed on the same plane: the principle and the accessory. The thought of the young child is not as structured as that of the adult: there is no chronological sequence, no precise relationship of cause and effect. His thought is isolated, fragmented, consisting of juxtaposed objects. His reasoning is transductive, that is to say, he proceeds by chains of associations uniting concepts without logical necessity, perceiving analogies without controlling them.

But the child's powers of recognition can sometimes be superior to those of an adult: a syncretic grasp of reality allows him to identify an object from a barely visible clue that escapes adults. The fantastical range of these clues will be, the important parameter to

[68] DRĂGĂNESCU Mihai, *L'Universalité Ontologique de l'Information*. Available at: http://www.racai.ro/external/static/bookdraganescu/tdm.html.

[69] ERHRENZWEIG Anton, *L'ordre caché de l'art, Essai sur la psychologie de l'imagination artistique*, Paris: Gallimard, 2005.

consider[70]. It is this mode of syncretic thinking that can be put to work in the creative process: less structured than conscious thought, but providing a hidden order that does not appeal to the differentiation techniques[71] of reason. Moreover, the proper use of this thought, the unconscious faculties it develops, provides access to a very rich world of complex analysis and powerful creative output. *Unconscious scanning* is a state of consciousness that lets go of rational thought control, to permit the inferences that may then arise. It allows for a creative moment when facing the seeming chaos of future production, for integration of external and internal realities without going through the process of rationalization. Indeed, what may seem ambiguous, multi-evocative, or endless when taken on a conscious level, appears on an unconscious level as one more easily graspable structure. Giving access to an unfocused perception of the world, at once more comprehensive while also able to surpass in detail the accuracy of analytical thinking. It is both more flexible and in some cases more powerful than rational thought, as it has a wider radiation. This profound vision notably facilitates understanding those structures which are primarily deployed when offered numerous choices. The mathematician Jacques Hadamart speaks of this process when describing the need to globally but precisely grasp the structure of a mathematical argument, before being able to clearly see its different compositional steps. As Wittgenstein understood, we can grasp a word *in a flash*, while its precise meaning remains elusive and will

[70] Erhenzweig illustrates this fact in our ability to take three dots in a circle and interpret them as eyes and mouth in a face. Very slight modification of the relative position of these marks can change the physiognomic expression from sadness to joy, or even menace. Here, syncretic vision demonstrates a marked ability to recognize individual characteristics apart from any consideration of abstract details.

[71] Childhood syncretic vision does not differentiate the identity of a form when confronted with one of its details, but takes it all in together. It immediately grasps the total object, without considering abstract relations, like proportions, dimensions, etc.

appear only with its future use. The complexity of creative research, which must explore many paths, can benefit from this undifferentiated vision, without distinctions, more holistic. An area of thought where underlying ideas and underground affects intervene, below the surface screen maintained by consciousness. An expanded space where no law limits its contours, not even those that control linguistic expression.

Gaston Bachelard speaks of reverie as "the lucid dream and the rational abstraction, between the opaque unconscious and the translucent subconsciousness,"[72] a state of being, right in the middle of human knowledge, including the artefact subject to evaporation in the dream, or to condensation in objective knowledge. We believe emerging materials can, just like water, earth, and fire, allow our thoughts to follow amazing trajectories in a complex environment. The different characteristics of these emerging materials generate their own complex dynamics in the surrounding world. The fantasy anchoring these materials is essential to produce a discourse that is not fully preconceived and to more globally embrace a coherence which structures a much broader field.

Traffic[73] is an art installation that aims to project the material in motion, as if it were a movie. A circuit containing a fluid, the channels of which are the size of a hair, is located in a system of optical projection and the image magnified a hundred times. Small drops of another fluid, immiscible in the first, flow through the channels. Their

[72] ENTHOVEN Raphaël, "Gaston Bachelard, le dormeur éveillé 5/5 : l'Imagination", in *Les Nouveaux chemins de la connaissance*, France Culture radio program, 02.07.2010 ®C 10 00, available at:http://www.franceculture.fr/emission-les-nouveaux-chemins-de-la-connaissance-%E2%94%8209-10-gaston-bachelard-le-dormeur-eveille-55-l-ima.

[73] *Traffic*, Dominique Peysson, 2014. installation, art and microfluid science. Designed with Adrien Bonnerot. Scientific advise: Patrick Tabeling and Fabrice Monti, MMS Lab, ESPCi. With the backing of PSL*, EnsadLab and Paris 1 Panthéon Sorbonne.

shapes are perfect and their movements very regular because the fluid has the property of being laminar at these dimensions[74]. The transportation of drops takes place before our eyes, evoking in its consistency the visuals of the first computer games. The operations of division, coalescence, or even the inclusion of one drop into another following behind bring other lights onto these drops of fluid, making them appear primarily organic, a little like living cells. The system establishes a link between small and large projections by deliberately ignoring the digital for simple analogue principles. Changes in rate accompany changes in drops, sometimes developing a historical morphogenesis of the drop, sometimes a series of cellular divisions.

The analogies that *Traffic* inspires are essentially dynamic: the mechanics of clean fluids at such small size scales, at the limit between our macroscopic world and that of our cells, which govern the processes taking place before our eyes. Or other movements, such as those that develop throughout our body, taken as a whole, but which can be observed in its interior. White blood cells that swallow enemies scattered in our blood, the fertilization of an egg, the cells' development after fertilization... a story of an inner realm that is ours, but which we never have access to other than through still images. It is this dynamic material which is given to be seen, so everyone is able to generate possible analogies using his own inner movement or the principle of life, that of our cells or at a different scale than our own evolution. The possibilities are all the more open than the relationship between the image-movement provided and the projections we can make are not imposed by the image, but rather decontextualized.

[74] Some phenomena, such as the capillary, were negligible in macroscopy and then became paramount; conversely, other forces such as gravity became negligible. The flow is called laminary if it is stable and not turbulent.

Images of complexity

> True symbolism lies where the particular represents the universal not as a dream or a shadow, but as a living and passing revelation of the unfathomable.
> Johann Wolfgang von Goethe

The assumption that "thought produces knowledge when it creates an image of complexity," proposed by physicist Jorge Wagensberg[75], has as its corollary the fact that knowledge is necessarily finite, while complexity is, in principle, infinite. For Wagensberg, always, art is a form of knowledge that agrees to treat complexity, even when the project of scientific representation becomes unthinkable[76]. Then it is knowledge that does not necessarily require intelligibility of the totality of the object under study: it creates an image without support in that this image is independent of the mental processes of the artist, and without stating universal propositions. Each time it provides a finished image of an infinite complexity without imposing any limits whatsoever. A kind of projection according to a specific plan, constantly reiterated over other different plans, of a certain complexity. It does not become intelligible but *recuperable*. Art would be based on its ability to *represent* the idiosyncratic[77] thought of the artist, communicating with other thoughts its singular complexity at a particular point. A process that allows reconstruction of an original complexity in a syncretic manner. For Wagensbert, the ability of art to evoke mental images makes it a mode of knowing, the details of which are based on a single principle:

[75] WAGENSBERG Jorge, « Idées pour une imagination impure », in *Percolations*. Paris : PUV, 1998, n°16. (Théorie Littérature Enseignement).
[76] WAGENSBERG Jorge, *L'âme de la méduse. Idées sur la complexité du monde*. Paris : Seuil, 1997, p. 107-141.
[77] Characteristic of an individual, which distinguishes by its manner of being, its behaviour, its modes of reaction.

"the principle of the communicability of unintelligible complexities.[78]" Wagensberg takes to use as an example intelligible complex truth and passionate love, which can be felt but cannot be rationally described. The expression of this passion through the generation of a poem or a visual work can be a means of effectuating a simple projection, a finite testament to infinite joy. It is from this testimony that others can deduce an original psychological complexity because it activates in us, like a signal, our internal mechanisms similar to those the artist perceives in the throes of her passion.

The artist can take advantage of the innovative features of emerging materials and intuitively integrate them into his compositions. The degree of material complexity does not necessarily appear if the artist uses it wisely but does enable enormous expansion of the understanding of the possible and of imagination. An extension that multiplies the means of expressing the degree of complexity of our contemporary relationship to the world, our current strategies of representation in a hyper-technological world, but which is empty of meaning: the *extreme contemporary*[79].

Louis-José Lestocart also addresses the subject of aesthetics and complex thought. He does so with a very physiological description of the possible transmission of the complexity of a system for our perception[80], neuronal communicability. He speculates that mirror neurons, highlighted by neurologist and physiologist Giacomo

[78] *Ibid.*, p. 111.
[79] The extreme contemporary (the term proposed by Michel Chaillou in 198) defines literary production of the last twenty years. We borrow this expression to give it another sense, which closer to a thinking of extremes and a strong belief in the complexity of our contemporary world.
[80] LESTOCART Louis-José, "Théories de la prédiction", in *L'intelligible connaissance esthétique*. Leonardo/Olats, décembre 2008, p. 17.

Rizzolatti[81], allow interaction between complex systems and the one who perceives them. The operating modes of mirror neurons, characterized by their complex processes (feedback loops, stochastic threshold effects, etc.) make this direct communication particularly effective. These neurons, coupled with traditional neurons associated with action, create an evolving and elaborate visual-spatial mechanism, able to reproduce mentally, and therefore virtually, the action or movement envisioned. They can therefore produce a phenomenon of resonance between the action perceived through the work and a virtual actuation in our mental field, of intersubjective imitation. Also, according Lestocart, complex processes internal to or depicted in the work could be coupled to similar phenomena taking place in our brains. The meeting of the two would lead to the "aesthetic feeling" felt by the observer.

Physical phenomena in emerging materials, which develop for the most part their own dynamics to complex phenomena, can in this way access synchronization with our intersubjective mental field. The deep structures, perceived via intensities, fixtures, reliefs and depths, and movements, can immediately communicate. Emergent properties, derived from a cooperativity of the parties, and feedbacks that connect cause and effect and self-reference, undecidability, and incompleteness, are all characteristic of these materials. The nonlinear phenomena are, also manifest, leading, by example, to qualitative threshold effects: new features appear from a certain value of a critical parameter, like a chaotic state from a stable condition.

[81] RIZZOLATTI Giacomo, SINIGAGLIA Corrado, *Les neurones miroirs*. Paris : Odile Jacob, 2008.

Living systems involve the same type of dynamic processes. Emergence happens within the need to ensure continuity (for reproduction) and requires circulation of flow and exchange. Families may develop, grow, interact, and then die according to the primitive behaviours of life. The parties form a totality that exists in a container, a kind of operational closure.

2. MATERIAL-POETRY

> "Imagination is not, as suggested by etymology, the ability to form images of reality; it is the ability to form images that exceed reality, which sings reality. It is a superhuman faculty.[82]"

Bachelard's *material imagination* come from substance. The material develops according to two distinct schemes, that of deepening and that of development. "In the sense of deepening, it appears unfathomable, as a mystery. In the sense of expansion, it appears as an inexhaustible force, a miracle.[83]" But, according to Bachelard, there is another imagining force, that he describes as formal, which comes alive in variety and vividness, in metamorphosis and exuberance. The imagination of emerging material mixes both modalities, sometimes quite intricately: while keeping its primary materiality, it is capable of transformation, of upheaval, of the extraordinary, all in the manner of formal imagination.

Almost thirty years after the *Immatériaux*, the exhibition-event at the Pompidou Centre curated by Jean-François Lyotard and Thierry Chaput, the materiality of work which was thought as immaterial at the dawn of postmodernity now permits another kind of materiality.

[82] BACHELARD Gaston, *Water and dreams*, op. cit., p.25
[83] *Ibid.,* p. 9.

The material according to the *immaterial*. A materiality that, in our post-post-modern era, cannot be only that it which was before, the paste come out of the tube or even the free resin of Niki de Saint Phalle's shapely sculptures. It takes note of its passage through immateriality to return – more complex, more thoughtful, more mysterious (less crude, finally) in the hands of artists who come to search for it the laboratories. It borrows the mental material of some relational patterns formed during the past thirty years between the work, the artist, and the public, while loudly and clearly claiming deep roots in the ground *that the hand knows*, its weight, its heart, sometimes its life. Form and content all at once, it is talkative and claims its imaginary...

The psychoanalyst Guy Lavallee tells an anecdote about Matisse. Commenting on the completion of one of his paintings, the artist explained how, wishing to paint lilies, he realized while painting that he had actually rendered clematis, flowers that were in the hedge of his garden at Issy and within him unknowingly. The painter represents reality but through this he also calls upon other images, those that affect him, most intimately. Doesn't the suggestive power and the song of the soul put forth by poets also draw upon a libidinal regression that links what we see to what we feel, or our past emotions? Are these other dimensions that come in layers, superimposed on our simple vision?

Progradience and regredience are two psychic movements described by some psychoanalysts[84] as opposing dynamics in our relationship to the world. The first is that of the act: it alerts us to read the world as it appears to us for our participation. It is the desire for

[84] Such as Michel Fain and Guy Lavallée.

future time, to realize our projects. It is the will of Matisse to paint the flowers of his painting. Regredience, contrarily, is introjective: it tends to formal regression of word to image. Receptive and passive, it turns to the past and aims for a return to a soothing calm. Regredience can be induced by a hallucinatory push, an image fully loaded with a multitude of other precepts and memories charged with affect. This strength comes to thicken our representation of the world, a kind of libidinal varnish that brings richness and depth. Without altering reality, this hallucinatory charge gives lifeness to oneself and one's world. The hallucination is the engine of hope, the energy that fuels the endless investment in speech.

Creating a poetic work demands that the artist place herself in this dialectic between progradience and regredience as the work does to the audience. The great poetic force of the materials of the future, like the lilies of Matisse, is that they carry in their folds the seeds of the hallucinatory and the anchor points of an active will.

A number of contemporary art works produced with emerging materials have been made using this unique ability to *image*, in the original sense of the term. This is particularly the case with the works *Hydrophily*[85] and *Color Metal*[86], which reveal patterns of fine and precise vapour composed of small water droplets on a transparent support. This is also true of certain works made from nanomaterials, like the *Starry Skies of Nanolithographies* by Alain and Laurence Ressier Josseau, or the work of Alessandro Scali and Robin Good, a

[85] Interactive work, 2013. Dominique Peysson and Hsinli Wang. Part of the series of works, "Les Limbes" by Hsinli Wang. With the support of EnsadLab, PSL

[86] Performance, 2013. Dominique Peysson and Hsinli Wang. After the series of works "Les Limbes" of Hsinli Wang. images of vapor appear as if by magic during the reading of a text, then the vapor is charged with more and more water, the image progressively coming apart and flowing. With the support of EnsadLab and PSL.

camel sculpture small enough to enter the eye of a needle. Works invented at the limit of life can also create an image; the algae-graphs of Lia Giraud are a good example. A population of green microalgae in a culture medium moves towards the light when the artist projects a photographic negative. Algae will take a break to image from the different light intensities : the more the negative lets in light, the more microalgae are focused, and the greener the colour. The *Arbrorigènes* by Ernst Pignon-Ernest are shaped by the artist before being delivered into their own lives. Several works, such as those of Eduardo Kac, *Specimen of Secrecy About Marvellous Discoveries* and Daro Montag, *This earth*, are themselves composed of the action of microorganisms left in the work, little by little becoming an entire tableau.

Some images appear immediately as metaphors of other strong images held in our imagination: images of small drops included in larger drops, kinds of cells-works, contain an analogy to cell fertilization.

What is the poetry of these images, both coming from the material and from receiving a more or less defined representational function? What is their power, their mystery, their authenticity, their tension, their presence? What essentially differentiates more traditional art images in their support (paintings, photographs) and digital virtual images? To answer these questions, we begin with an analysis proposed by photographer Henri Peyre about the contemporary photographic image[87], and get even closer with a quotation from Gaston Bachelard[88]. The principle is simple and Bachelard stated in

[87] PEYRE Henri, *Mécanique esthétique de l'image : théorie des mondes*. Galerie-photo [online]. 2011. Available at : http://www.galerie-photo.com/mecanique-esthetique-image.html.
[88] BACHELARD Gaston, *op. cit,* p. 19.

this way: for the material element to attach to the soul, it needs to have "double participation – participation of desire and fear, participation of good and evil, quiet participation of white and of black [...][89]". Henri Peyre, meanwhile, in the specific context of art photography, speaks of an enjoyment available when the image permits perception of several different views of things. It is this passage from one to the other, this precise moment of hinge, of tilt, between the visible and our interpretation, that causes enjoyment. Requiring both a world and an elsewhere. An enjoyment that takes its source, he said, from its similarity to the paroxysmal loss of control in sexual enjoyment: the tipping of our *individual consciousness* into the *consciousness of the species*; or, more generally, in switching from one world to another. A bipolarity in any event, capable of generating power as between the positive and negative poles of a battery.

It seems appropriate to use these presuppositions to talk about our image-material. The passage between two worlds, the double participation, can very often be played at the level of the material that embodies the image. It appears as evidence for the works considered as both the material and its mode of production, rather than disappearing behind the image, are stakeholders in the figuration. To view *Hydrophily* is to see the image of water plants rippling in ponds, the designs of the fog on the tank walls. But it is also, through a mental spillover, to see small drops of water assembled there and shine under the light. In reality, these droplets hang on the surface, as in the familiar image of dewdrops on morning leaves. So matching the water pearling on plants. But the inner eye can switch again, this time seeing the fog on the glass, as on our

[89] *Ibid.*

bathroom mirror from the moisture of the shower. A mist that reminds us of drawings traced by a fingertip across the glass, magic, tiny wet signs through which you might see the night. Small openings through which curious, but protected, we spy at the world outside. Yet another world! The moving light in the *Hydrophily* installation, takes in the droplets. And the water, that of rain, drops, streams, or ponds, also evokes a world tinted with nostalgia. All these worlds may be convened at the same time, each acting toward each other as does metaphor in a poem. Their power will be more intense the more extreme their comparisons – as long as they are still accurate – still well-founded. "The picture [for the poet Pierre Reverdy] is a pure creation of the mind... The more the relationship between the two realities drawn together is seen as distant but legitimate, the stronger the image will be, the greater its emotional power and poetic reality.[90]"

The material of water is often associated with the elapse of time: *Hydrophily* engages this directly with the drops which, when they become too large, flow along the glass. Mist materializes this loose veil over memory, which softens the features but can clear all at once and keeps a trace of its contacts. *Hydrophily* can therefore be seen as an attempt to hold back time, the instant, that of the ephemeral, to keep it forever. The material here is *deepened* and mysterious, as Bachelard suggested. "Life is a vapour, light mist that dampens the palm of the hand and then nothing.[91]"

[90] REVERDY Pierre, *Le Gant de crin*, Paris: Plon, 1927.
[91] Excerpt from the poem, *Celui-là était roi*, by the Mexican poet Jaime Garcia Terres.

a- 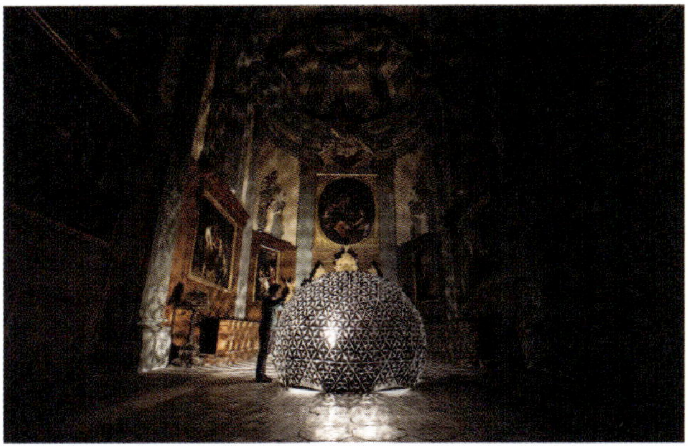

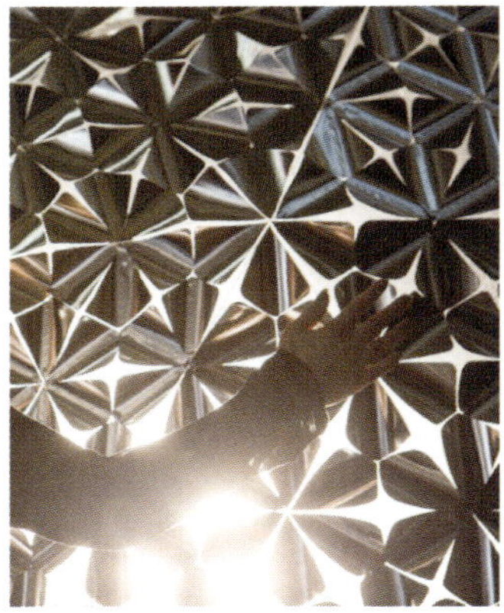 b-

4. Memory-matter

a- *Lotus Dôme*. The installation by Daan Roosegaarde / studio Roosegaarde 2010-2011 is made of thousands of small sheets of shape memory metal that folds when you approach by creating an interactive relationship of light and shadow.

b- *Lotus Dome*. Detail.

c- Artificial muscle for an android robot made using shape memory alloy. A muscle fibre 100 times stronger than the biological muscle obtained by researchers at MIT: Eduardo Torres-Jara, Kyle Gilpin, Jonathan Karges and Robert Wood.

d- *Hylozoic Series, Sibyl*. The site-specific installation by Philip Beesley, Sydney, 2012, refers to "hylozoism" an ancient belief that all matter has life. Here, the meshing of touch sensors creates waves of diffuse breathing movements designed to draw visitors into the depths of a forest of light. These creations are the hybrids of sculpture, engineering, chemistry, and experimental architecture able to "feel", "to be empathetic." @ PBAI (Philip Beesley Architect Inc.).

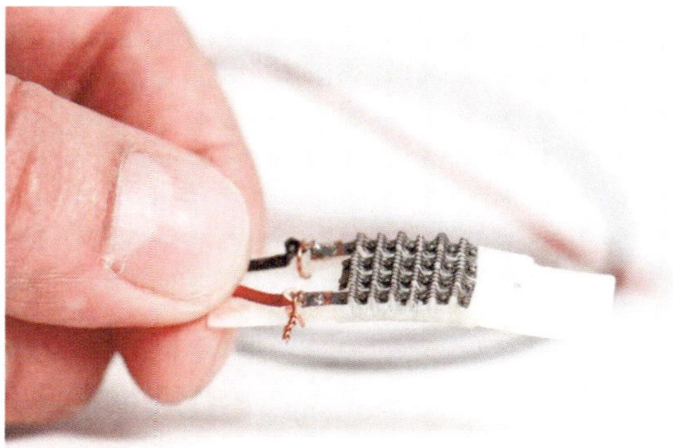 c-

d- 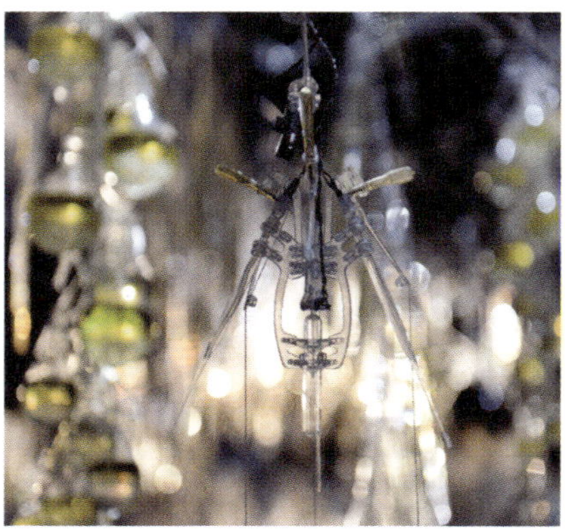

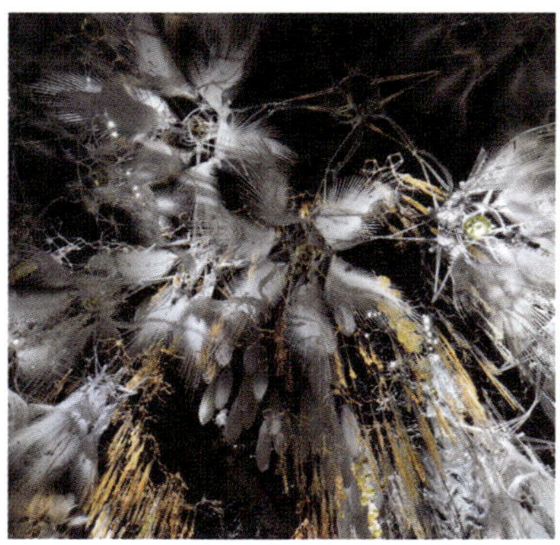

The intense presence of absence

Hans Belting's *An anthropology of images*[92] evokes this function of image as the double of the body, replacing it where it is absent, facing both the disappearance of the other and the next disappearance of the self. Studies of the first psychic images of childhood shows that this function is essential for each of us, since it allows us to form psychic image of the one who is physically absent, providing him or her to us as mentally present[93]. The least of these separations, that of baby from mother, is better endured if an internal image representing her keeps a little of her inside us.

In fact, what makes the image unique is that it leads us to believe that it has unlimited powers; reaching for us a sort of absolute that it alone can approach : "the site of an exceptional marriage of the visible and the invisible.[94]"

Mental images inspired by emerging materials benefit from an intensification of this power of representation that adheres to all image, the presence of absence. Because of their strong presence in the world, these functional materials respond to their environment in an original and often very powerful manner. *Cognitive dissonance*[95] can result from this tension between our perception and our expectation of the material, and may be put in service of the image to intensify the feeling of presence.

The intensity of these images can cause remanence – the term remanence typically characterizing the persistent capacity of a phenomenon or sensation after the disappearance of what caused

[92] BELTING Hans, *Pour une anthropologie des images*, Paris: Gallimard, 2004.
[93] TISSERON Serge, *Psychanalyse de l'image*, *op. cit.*, p. 217.
[94] *Ibid.*, p. 41
[95] BORILLO Mario, *Dans l'atelier de l'art. Expériences cognitives*. Seyssel: Champ Vallon, 2010, p.59

them. This persistence, characteristic of certain physical quantities, can be applied here to our mental images. The stronger the emotional intensity – and emerging materials contribute to this – the more effective this remanence will be, enabling us to preserve this presence of absence.

The defamiliarization of the image that may be obtained with emerging materials because of their singular features participates in the *mana* of the image, to differentiate between its omnipotence and the common familiar objects around it. Contradicting those texts which foresee a life further from the material, which follows the formal logic of technology, which would displace the sensual in favour of the intelligible, the high-tech of emerging materials bypasses the pitfalls of this distorted experience. Unlike a photographic image or a digital image, reproducible countless times, the unique material image of which we speak, draws from the perceptual intensity of what makes its presence to the world its capacity *to be*, intensely. Its strength of transference between presence and absence. Its willingness to remember, for retention. The loss of the *aura* that Walter Benjamin[96] spoke of no longer applies: no mass reproduction, no work disembodied due to its reproducibility and its declension in infinite sub-models. The image is inimitable, the carrier of a singular vibration. It speaks several languages: that of the visible, the touch, and finally, the body.

And yet this capacity for a strong presence must not be at the expense of a sense of meaning or of the symbolic. Paul Ardenne, again, speaks about the meaning of the image in art that cedes to "the effect, the spectacle, and the attraction[97]", the allure of a line of

[96] BENJAMIN Walter, *l'Œuvre d'art à l'époque de sa reproductibilité technique,* (1939) Paris : coll. Petite Bibliothèque Payot, 2013

[97] ARDENNE Paul, "L'image d'art contemporaine: impossible définition et stratégies de recomposition". *L'art même,* n°27, p. 4. Available online:
http://www2.cfwb.be/lartmeme/no027/pages/page4.htm.

cosmetics that would be triumphant today. At this point, it is certainly necessary to be vigilant. Art, according Ardenne, is of course a simulacrum, which does not a priori limit its potential depth and authenticity. Provided that this simulacrum does not act to the detriment of the meaning of the work, only to be constituted as display of fireworks. Leading to the very essence of a fireworks show in works like *Fireflies*, *Waterlilies* and *Snowballs* by Bruce Munro: works that certainly delight the senses, both for the artist-artificer and the viewer. Able to trigger pleasure but nothing more. Losing the power to intervene on thought, on the ability to carry conviction, "without anything in terms of meaning firmly rooted or coming together.[98]" Excellence has given way to availability, adaptation of supply to demand, the eager response to social desire. The temptation is obviously *to have an effect*, to play with surprise and engage the machine *even more*, which is what emerging materials know to do so well: to be higher, stronger, bigger, more difficult, more surprising. And of course, more beautiful.

The challenge is this: to avoid the lone fast-track of the superhighway and prefer a path through the fields, trampling the material underfoot, taking its measure, stringing it to encounters with another. The latter can also do many other things that make it beautiful: it must (it can) be a little bit deeper, a little bit heavier as well. Having consistency, a sedimentary heft. Or, conversely, a light vapour, closer to ether. Blurred (why not), cloudy, ambiguous. Intrigue, in the way of presenting a plot, a story, of posing questions without answers. Troubling our precepts and thus, the real. Giving way to indecision about what we see, initiating a movement of thought but leaving it to the public to discover its wake. To give them the responsibility to

[98] *Ibid.*

construct a sense, of wanting to see things in perspective. The fog, for example, is ghostly, goes away and comes back, is *inframince*, as Duchamp would have said.

Works of art also give us beautiful demonstrations of the aura that can be born by images generated by emerging materials. Another natural material, unique in its structure, is the gold in Stefane Perraud's *The Rule of Three + 1*. In this work, the artist restores the arms of a wood and ivory crucifix with a moulded polymer, which he then re-covers in gold. The structural state of the gold differs for each sculpture: nanoparticles of pink gold, of blue gold[99], and of gold leaf. Three colours for one element, gold (Au), symbol of immutability. Here, the artist works directly on the most powerful religious icon like a medieval alchemist magically transforming lead into gold. A strong entry in the histories of both religion and magic. Allowing us to say, as for certain other works, that the emerging material effectuates an encounter with a state of being, with wonder, with the magic moment, both here and now, above and beyond, far from other realities – all while inscribed in a period, as testimony.

This does not mean of course that all works made of emerging materials participate in this lineage. But we can share the secrets of composition of those, as André Breton said of surrealist art, that "re-engender somehow the magic that created it." Such as life, in the magic of *Algae-graphies* and *Temps-de-pose* by Lia Giraud, or, again, Eduardo Kac's *Specimen of Secrecy About Marvellous Discoveries*, and the secret action of creative microorganisms. The work *Or* by Marie-Julie Bourgeois, which stages a levitating coin,

[99] These two states of gold, pink and blue, come from synthesis techniques that create various sizes of agglomorations of nanoparticles, leading to different colours. It acts here as a physical colours due to the diffraction of light, like that of butterfly wings.

a-

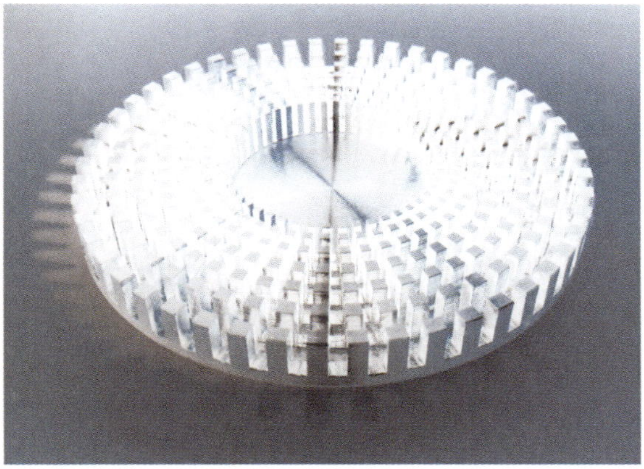

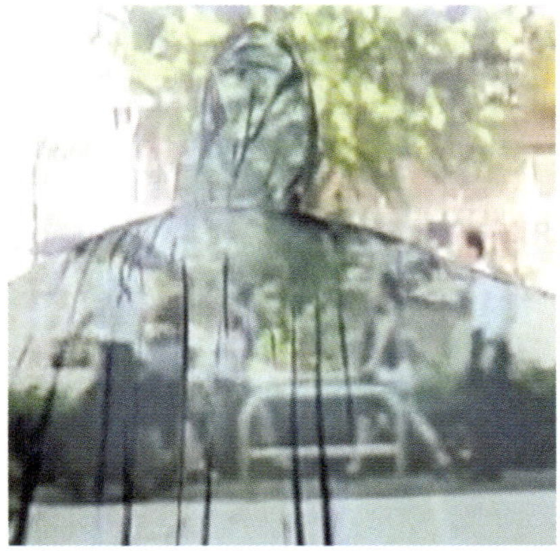

b-

c-

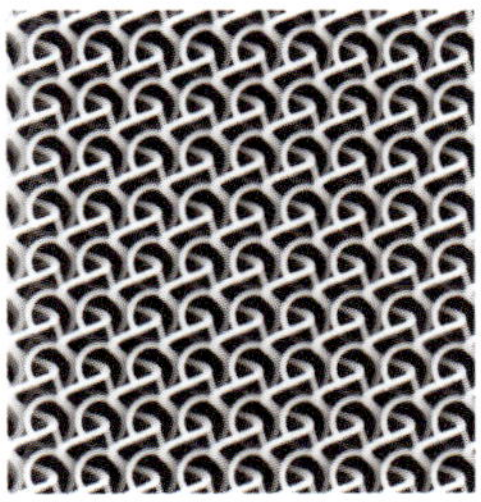

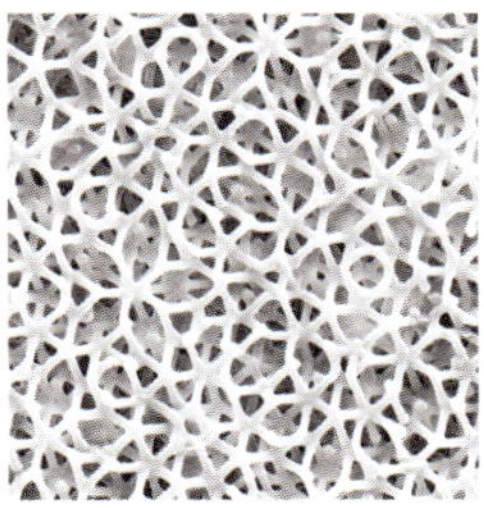

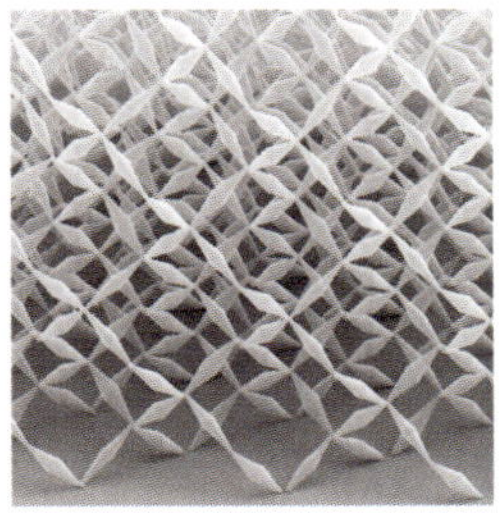

5. Infrastructure

a- Metamaterials: which render objects invisible; here creating an anti-tsunami shield. It's the complex internal structure of the material, produced in the lab, that allows it to deflect waves. Metamaterials offer varied and unexpected paths for applications, such as neutralizing waves, protecting against earthquakes or becoming invisible. A material that does not exist in nature. Photo S.Enoch / CNRS photo library.

b- An invisibility cloak like that of the apprentice wizard Harry Potter. Researchers at the University of California, Berkeley have fabricated a prototype able to hide a small object with nano-antennas that redirect light, rendering the object optically undetectable.

c- Insect wings present structures called photonics that control the propagation of light. Physicists have been inspired to develop new materials. Here there are three microstructures: a bichiral photonic crystal, a photonic quasicrystal, and a metamaterial pentamode. @Joachim Fischer, Michael Thiel & Martin Wegener.

d- An ultra light material consructed from 99.99% air but ultra durable ; the secret lies in its structure. A micro-lattice made from a liquid polymer hardened under UV then covered with metal, constituted by air for the most part ©HRL Laboratories, LLC/Photo by Dan Little.

e- Aerogel graphene has a spongy structure weighing only 0.16 mg / cm3 but able to absorb up to 900 times its own weight in petrol, oil or water, and then release them. The researchers at the chinese University of Zhejiang used a novel freeze-drying method.

d-

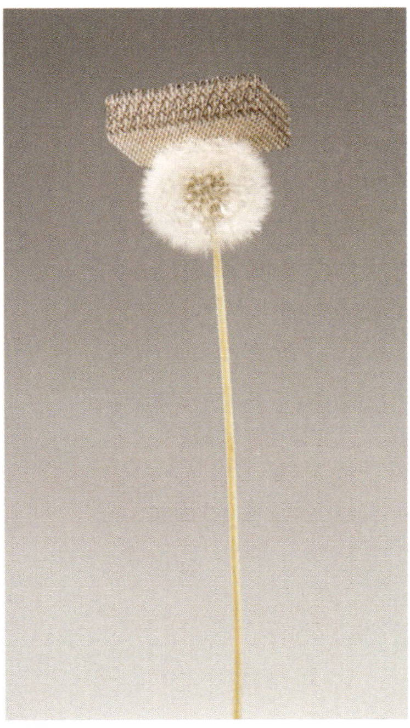

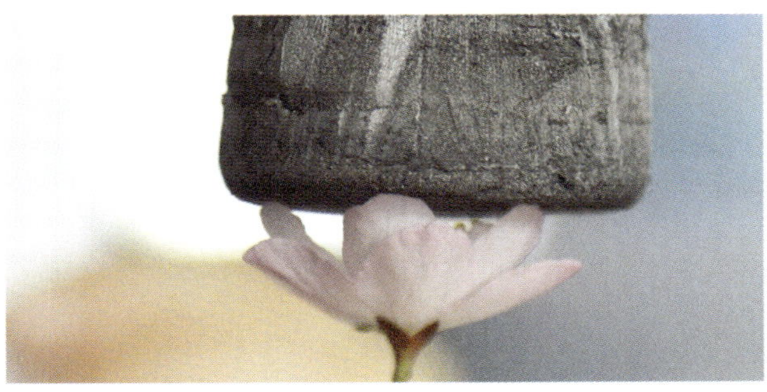

e-

oscillating between two states (rather than falling heads or tails) fits well in our story. The coin is kept suspended in the air due to forces exerted by a superconductor located beneath. The installation is certainly technical and very difficult to achieve but the technology is effaced to make way for the magic of levitation, for the suspension of time (or at least gravitational forces). The work makes an instrument of observation: attention is directed towards the forces that seem to have pulled free from the singular object. And thus "emerges the distinction between figure and ground in the universe.[100]" A powerful imbrication of space and time, "separated from other realities by an irreducible distance[101]", a work that certainly has the aura of a unique art object. Paul Thomas's *Nanoessence* is an audio-visual system that permits interacting with the image of a human skin cell, magnified to the scale of the atom by an atomic force microscope (AFM), and which can be seen dead or alive. Thomas sets the limits of a single skin cell's life and death to interrogate the concept of life at this scale. A technical resolution to the question of the transitional passage to the after-life, after-death, usually taken up by religion.

Becoming poetry

For Umberto Eco, the aesthetic pleasure aroused by a work of art lies less in recognition of a stable form than in following the process of its opening.

The peculiarity of emerging materials is... precisely in being *particular*. This material uniqueness, which can be surprising or confusing, escapes ready-made representation. Our thinking can play

[100] MICHAUD Yves, "Gilbert Simondon (2)", Philosophie Magazine, janvier 2013, Available at : http://www.philomag.com/blogs/philosopher/gilbert-simondon-2.

[101] BONTEMS Vincent, "Quelques éléments pour une épistémologie des relations d'échelle chez Gilbert Simondon", in Appareil [on line]. Septembre 2008 vol.2. Available at: http://revues.mshparisnord.org/appareil.

in this in-between of the reference to the known and the unexpected or unknown. Human thought characteristically builds gradually by referring to the *already known* and categorization by analogy[102]. It is a mode of understanding our reality that we see at work in the words of children, as with the little girl, Clementine, who asks her parents in her bedroom to turn down the music in the living room, saying "Softer: it overruns !" Her representation of sound is closer to an *invisible cloud* or a material capable of invading more or less space. The absence of a representative model of the physics of sound led her to more familiar images better known.

Many emerging materials activate physical processes which are surprising because they are different from those we encounter in our daily lives and for which we have no readymade analogy. This is the case with Ferrofluids, shape memory alloys, intelligent fabrics, bioluminescent organisms, etc. *Or*, on a different level, the magical appearances of images in fog on glass, the movements of small systems of non-motorized wood, the image created by tiny algae : in short, systems we know well but which have adopted modes of operation that seem unnatural to us and whose behaviour we can no longer predict. The multiplicity of mental interpretations that each build will be enormous. The images summoned by our need to understand the functioning of what surrounds us will be very different for every viewer and depend also on the form of the work given by the artist. Working on the imaginary, the artist takes advantage of the greater manoeuvrability of the materials' power of evocation. The displacement of small balls of quills on a smooth surface of a Ferrofluid is very quickly assimilated by analogy to that of extraordinary little animals (as yet unrecognised). This

[102] HOFSTADTER Douglas, SANDER Emmanuel, *op. cit.*, p.26-40

colourful projection of animal life would probably not occur to us if we were used to seeing these special fluids and their prickly expressions.

If we consider the work of Marie-Julie Bourgeois, *Or*, it is evident that the levitating coin directly refers, in varying degrees, to many more or less implicit meanings, as well as having a powerful emotional potential. Obviously, the coin alludes to both money and time. But to a degree, we can also say that, more than the object, it's the situation of this object caught mid-flight that matters. It's the suspension of the coin and its oscillation between heads and tails, whose strong signification resonates in our collective imagination: that of a game of chance, surely, where everything is decided, of the transitional passage towards the inexorability of what will come to pass. Of determinism, as well. Using a coin reinforces the range of the outcomes of this game: on one side wealth, success, happiness. The other… misfortune. The end? An "emotional halo" results in any case, highly suggestive, of very different kinds, depending on each person and their past experiences, more or less positive. The metaphor, that of time, of chance, of searching for control over the imponderable events of life, of the fragility of our destinies when everything can change in a heartbeat, is in fact an emotional use of reference, or the referentiality of emotions. Anchored in specific media, the emotional orients and amplifies our perception of the senses. The work, here the levitating coin, allows identification of sign and signified, a sign that qualifies, according Eco, as iconic[103].

[103] ECO Umberto, *L'œuvre ouverte,* Paris : le Seuil (coll. Points Essais), 1965, p. 55.. The aesthetic sign frequently has this feature of being iconic and designating a value. Because it is a sensed form, the manner in which it is presented will account for the aesthetic.

Feature-rhythms

The relationships we have with the world are in first place temporal, as Serge Tisseron reminds us: "Some specific psychic structures [that] ground the image as a possible place of transformation, of the self or of the world [...] are organized at first by rhythms, which are the initial mode of apprehending perceptual experience.[104]" While digital installations can force time into any kind of countdown timer or imposed break, the material here can only play its own score.

Systems that put materiality to the test may be evolutionary, under (or out of) control, stable, out of equilibrium or metastable, convertible. All these developments follow the materials' own rhythms, the rhythm necessary for poetic construction. Because the structures and properties of matter will induce specific dynamics that intimately correspond to them. They have what is called in physics their *characteristic time*, their *resonant frequencies*, their *time spans* for periodic systems. Their rhythms, which combine together to participate in their poetry. Any physical system, as soon as it is moved by the artist from its position of equilibrium, will find ways to regain the stable state to which it aspires. It is this inexorable search for balance and strength which directs it, that will enable the material's poetic rhythm to flourish.

As the analyst detects in his patients a propensity for repetition that forces them to reproduce – unconsciously – the same conditions ever leading to dead ends, so as to help them see their process and be able to extract themselves, the installation draws this arrow of time before our eyes, to see it from a distance and take its

[104] TISSERON Serge, *Le bonheur dans l'image*, Paris: Éd. Institut Synthélabo, 1996, p. 123.

measure. Extraction would be difficult, but the arrow may be played in any event. Given some slack, surely. The rhythms of the material structure and delimit the time characteristic of the system. They resonate with this distinctive time which mobilizes us physically. Or mentally. Sometimes faster, responding to our desires for animation and synchronization, desires (never completed) for achievement, sometimes slower, which evokes the cycle of all life: erosion or aging, the march towards death. A cycle which may be visited without staying there, in order to return.

Material to deal with anxiety

"Being an artist is using one's own anxieties,[105]" says Christian Boltanski, which keeps its traces to preserve their memory, his own, and that of others, as if he could prevent death. Mike Kelley, often seen as continually working with his own childhood trauma, built its architecture in the form of a model. Attempting through this physical structure to represent the places where he suffered from institutional psychological abuse. These are embodied through white blocks, undefined in the model, specific locations where trauma caused his memory loss. Kelly creates performances and other poetic works to fill the gaps left by the ordeal. "The idea is that you cannot remember anything, that you forget or block out, is the byproductby-product of abuse ; and that all of these scenarios are supposed to be filling in the missing action in these blank sections in this building.[106]"

[105] COLONNA-CESARI Annick, BOLTANSKI Christian, "Etre artiste, c'est utiliser ses propres angoisses", L'Express-Culture-Art [online], January 13th 2010 [accessed on October 7 2014]. Available at: http://www.lexpress.fr/culture/art/boltanski-etre-artiste-c-est-utiliser-ses-propres-angoisses_841605.html.

[106] "L'idée, c'est que tout ce dont vous ne vous souvenez pas, que vous oubliez ou que vous vous masquez, est le sous-produit de la violence; et que tous ces scénarios sont censés recréer des actions dans les sections de ce bâtiment qui en sont privées", KELLEY Mike, "Mike Kelley: 'Day Is Done'", in *ART21* [online]. November 2011 [accessed on 7 October 2014]. Available at: http://www.art21.org/texts/mike-kelley/interview-mike-kelley-day-is-done.

The work of art here is invested with a role similar to that of the psychic image. That is to say, it provides us with that intermediate representation, between the frighteningly unrepresentable and rational thought. The mental image certainly establishes a link between representation, our senses, and our bodies. It always has a sensory involvement and a strong participatory drive. Images borne by materials capable of articulating these inner storms, these distortions pulling towards the unrepresentable, of bodies torn apart, are mental images certainly capable of conveying feelings not yet secured, too violent, perhaps denied. The work of letting go, of *unconscious scanning*[107] described by Ehrenzweig[108], should allow the *empty-full* that we need in order to install our imaginings. It is in this outer field that the links between these material representations and our feelings can be established, in the original activity where the body and the organization of the sensory condition representation. Rapport may be established between the material – that of these complex materials – and the body. These are the depths of our sensible world which to be put to the test. The works do not evacuate the unfathomable of the body's felt. But they do reinvent a scenario that could play the part of protection against anxiety, such as with separation anxiety. The principle of an *inside* which can be *contained* is deeply reassuring for those who suffer primal anxieties. An envelope, an artificial skin that can protect them physically and psychically. Like artist Jalila Essaidi's dreams of hybridizing skin with spider silk securing its own psychic envelope. An artificial construct that constantly renews our condition as fragile and totally dependent infants, iwith the constant need to be reassured about the aggression around us and our own internal integrity.

[107] Unconscious scanning is an unconscious reflection that secretly analyzes a multitude of choices that would defeat a conscious search.
[108] ERHRENZWEIG Anton, *op. cit.*

a-

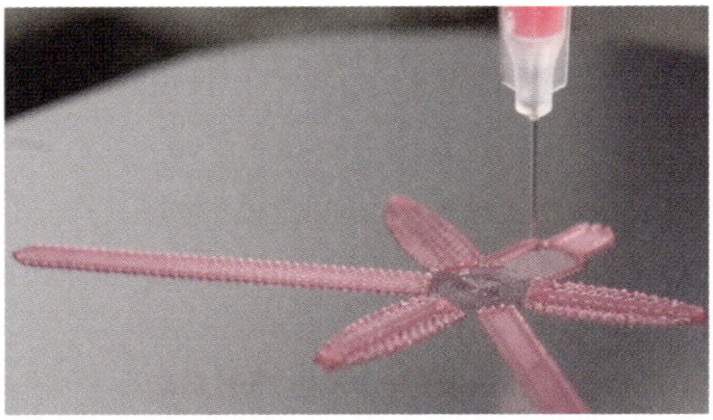

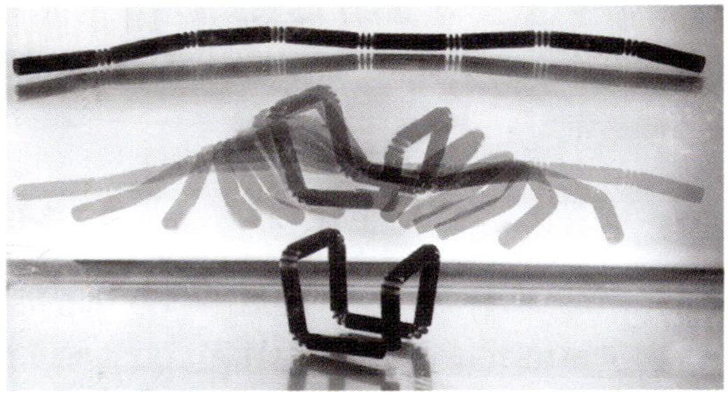

b-

6. Animated matter

a- 4D Printing. Layered structure (bilayer): one layer swells in the presence of water, the other does not, allowing for fabrication of material that become animated when the humidity changes. Here is an impression by scientists from Harvard Wyss Institute (United States).

b- A 4D material made by Skylar Tibbits, MIT. It is the structure of the material that induces the typology of its movement when placed in water.

c- *Dancing in the rain. Little ballet of papers*. Video. Dominique Peysson. 2014. Bilayer structures executing after one another a little ballet when sprayed with water.

d- Bilayers of plywood wood, as used by the architect Achim Menges.

e *ShapeShift* is an electroactive surface for architecture which comes alive through application of an electric field. It consists of conductive layers enclosing a dielectric layer. Developed at the ETHZ and EMPA in Zurich, Switzerland.

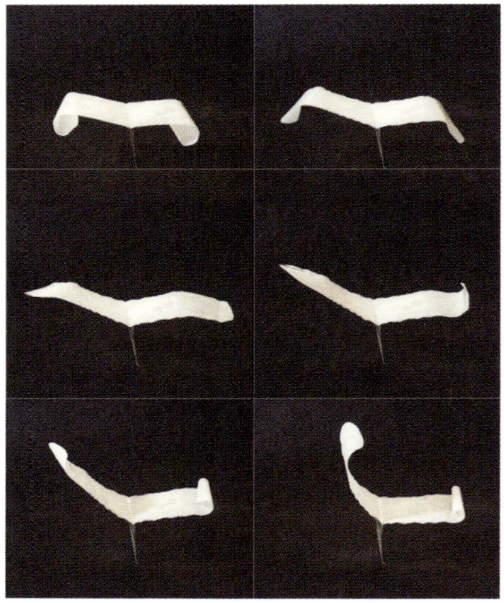

c-

d-

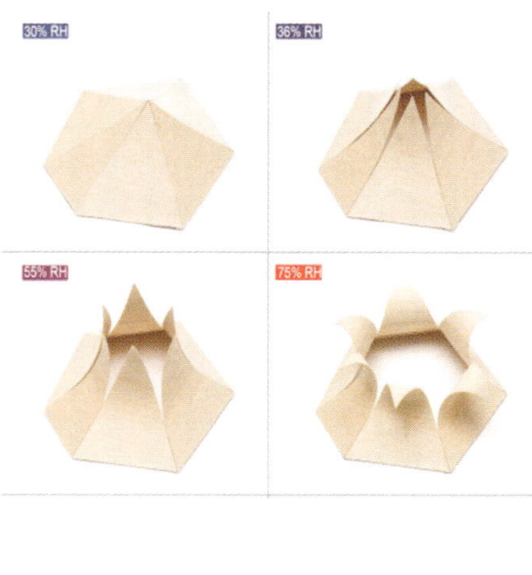

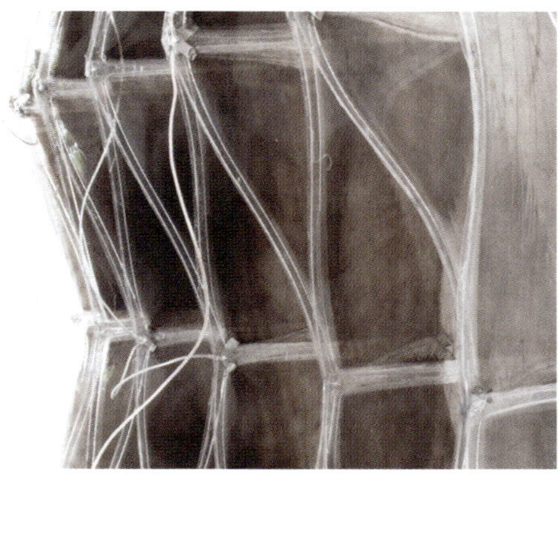

e-

These works – such as Kelley's model – are a way for the artist to produce a *representation* that is under his *control*. A real that is artificial. For the psychoanalyst-mathematician Daniel Sibony, art does not renounce the real but attacks the real of fantasy and all our interior realities that go unnoticed. It gives us a glimpse of strange realities that refer us back to our own strangeness[109]. The artist creates another world for the audience, a *surreal*, structured in its own way. For Paul Lawrence Assoun, the artist draws on his own background to "make something out of its material,[110]" if anyone fantasizes – to support his reality – the artist, himself, "acts" his fantasy in "work". Better: the artist effectively creates, to the extent that he claims to update something new, never seen, something innovative, as his work – manifest as an object in the world, coming from an outer-world[111]. Assoun described the work as the "improbable emergence of an 'off-world' at the heart of the world[112]". Emerging materials can take part in the manufacture of the *surreal*. Being themselves *sur-real* materials such as superconductors, super-hydrophilies, super-insolators, super-elastics – that do not exist in nature (metamaterials), they allow passage into another world, via links that bind them to our world. A surreal that is superimposed on reality but which has the advantage of being controlled.

For Dewey, the work of art can evoke and emphasize "the ability to be a whole, of belonging to a greater whole that includes everything and that is none other than the universe in which we

[109] SIBONY Daniel, *Création. Essai sur l'art contemporain*. Paris: Seuil, p. 9-11.

[110] ASSOUN Paul-Laurent, "Prolégomènes à une pensée analytique de la création" [online], in *Le Carnet/Psy*, 2014, n° 60, p. 20-22..

[111] ASSOUN Paul-Laurent, « La "création" à l'épreuve de la métapsychologie. L'objet inconscient de la création", *op. cit.*, p. 29.

[112] *Ibid.*, p. 33.

live.[113]" Aesthetic experience is then "the result of our active insertion in the world", an "experience of all" or an "aesthetic moment as an *open* moment of consciousness.[114]" Bachelard meanwhile, in the poetics of space, talks about the natural *imagination* that allows possession of the world as soon as it can be seen it in miniature. The miniscule is one form of the anthropological structure of our imagination, as emphasized by Marina Maestrutti[115]: a reverie of the miniature, referring to the dimension of the hidden, of the search, of the descent "into the nocturnal regime of the image" defined by Gilbert Durand[116].

Just as the study of the physics of these materials is not so much the study of their component elements but rather the structural relationships between the elements and the transformative matrices acting on these structures, what counts in an artistic work will also be the fiction of the world that it generates (its appearance, whether pleasant, disturbing or any other feeling generated by these creatures of the plastic imaginary), the transformations that will be accessed. The relationships that move us from the real to the sur-real of the work.

Practicing works involving responsive materials can distance anguish in a different way than that of a simple visual representation. It no longer acts only as an image-symbol capable of focusing, or at least circumscribing, anxiety. But refers back to this image in order to replay it in similar fashion, once, twice, again and again. Because repetition in a safe environment can be a mode of assimilation of the situational anxiety evoked by the image. This is the psychoanalytic principle of the repetition compulsion. Freud described the game of the spool, or the

[113] DEWEY John, *L'Art comme expérience*, Paris: Folio, 2010, (chapter 9).
[114] Eco Umberto, *op. cit.*, p. 47.
[115] MAESTRUTTI Marina, "Rendre visible l'invisible", in *Alliage* [on line], September 26-2012 [accessed on October 5-2014], n°70, p. 22, available at: http://revel.unice.fr/alliage/index.html?id=4058.
[116] DURAND Gilbert, *Les Structures anthropologiques de l'imaginaire*, Paris : PUF, 1960.

fort-da. Perfectly illustrating our proposition: a child repeating the trauma of his mother's departure through the game, despite the displeasure that the repetition causes. This isn't a real departure but a symbolic representation that the little boy enacts by dropping a spool fitted with a thread out of bed, only to be wound back. He accompanies this small scenography with *Fort – Da* (away – here), lending his name to this little story. This play enables the young boy, Ernest, to master his trauma, to appropriate it, to symbolize. This leads us to believe that the artist, not necessarily consciously, can intuitively produce a repetition of a revisited artificial reality. Similar to the original situation but not too much to be unacceptable. Like the spool and the thread, it contains its own means of control, which the viewer can seize rather than having it be imposed, and replay the comfort of a return after a separation.

Artificial living

What is this germ of order in the material that is capable of initiating life in the middle of chaos? At what point can we say that this assemblage of molecules that form a fully closed whole is a living cell? Scientists are racing us towards the Creator. Outside of the artificial of the virtual and auto-intelligent entities, arenas in which artists heavily invest, research on artificial life is largely artistically ignored. Meanwhile, artificial organs developed by doctors and other researchers in artificial life continue to set new standards. An obstacle course that we follow in the media, while knowing that we shall probably not see its end by the end of our lives, unless this puts them into overtime... To the point of displacing the line, still undefined, despite all our natural intelligent sciences, that separates *living* and *inert*, *natural* and *artificial*. And that ultimate question: what is life ?

In order not to get carried away as by a driverless car or emergent autopilot, society must produce a clear framework on how to act. Here, artists can help us represent to ourselves a multiplicity of imaginary extrapolations without inducing judgment. After all, re-presenting the real or natural through artifice is one of the principles of art. Reversing the process is also possible: re-presenting the artificial through reality. A typical example is the work of artist Jalila Essaidi and her attempts at metamorphosis, trying to unite in one tissue the softness of human skin and the delicate strength of spider silk. Or the work of American artist Julia Reodica, whose *HymNext Project*, cultivates human tissue in the lab and grows synthetic, and therefore artificial, hymens, from its own cells. The freedom enjoyed by art, born of this contradiction of not being useful while not being superfluous, clears a place to protect those places where financial interests flourish. Certainly a meeting point between philosophy and art, between bioaesthetics and bioethics, retaining of course a plurality of approaches to better understand complexity.

3. META-MORPHOSIS

If painting that captures and imprisons form by reducing it to two dimensions can seem essentially magical, the act of the sculptor is at once simpler and more daring: he creates the form, he is equal to the Maker who models man from the dust of the earth. Hence the secret desire to animate this form. Jean René Gaborit

One of our main hypothesis is that emerging material can give form to interactive systems of another kind: responsives. Putting intersubjective relations into play may occur intrinsically as the direct

a-

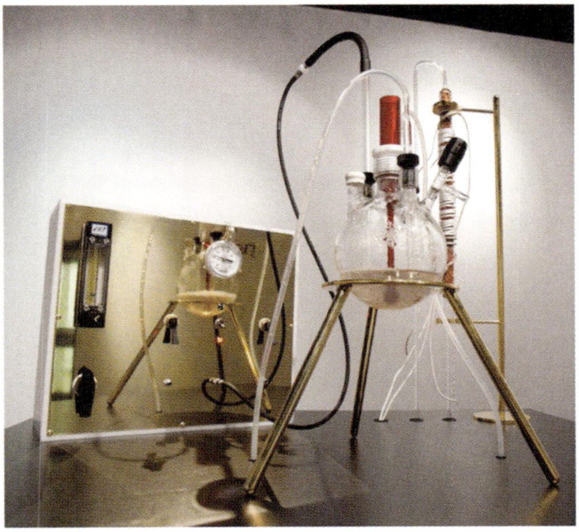

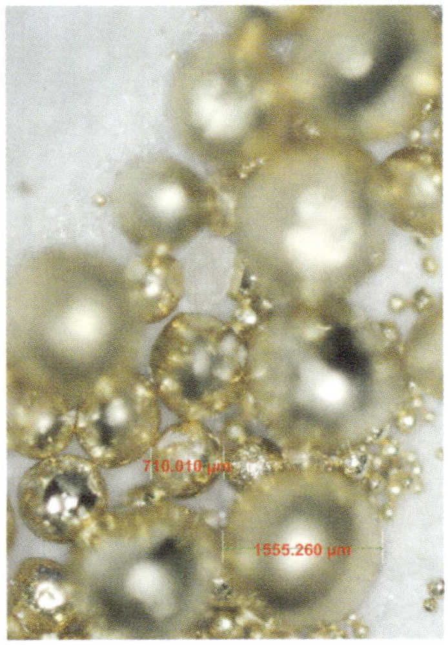

b-

c-

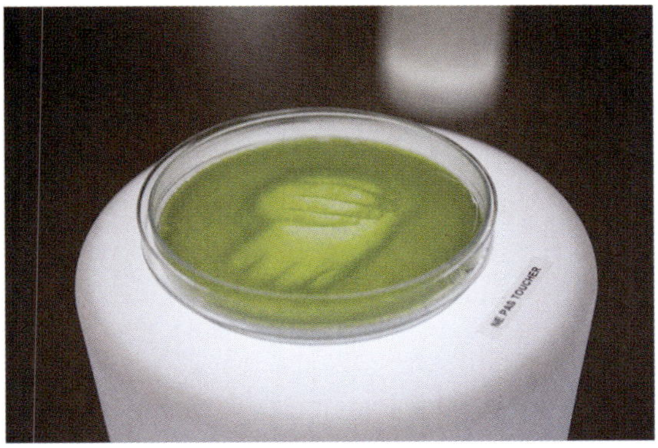

7. Vegetative matter

a- *The Great Work of the Metal Lover,* Installation, Adam W. Brown, in collaboration with Kazem Kashefi, 2012. Extremophile bacteria (very specifically tolerant to toxic metal environments) are able to metabolize gold chloride (AuCl 3) to crystallize pure gold. Presented here in a bioreactor.

b- *The Great Work of the Metal Lover*. Detail. The 24 carat gold obtained after extraction by bacteria.

c- *Immersion*. Installation. 2014. Lia Giraud and Alexis Raphelis. Lia Giraud has developed a technique to gradually produce images using live algae, sensitive to light, by projecting a negative image on a culture.

d- Yamanaka Furniture. Phil Ross. Cellulose is infused with fungi, providing them with both support and food. The fungal tissue is then dried and processed, becoming alightweight material that adapts to any shape. Photo Gregory Thomas.

d-

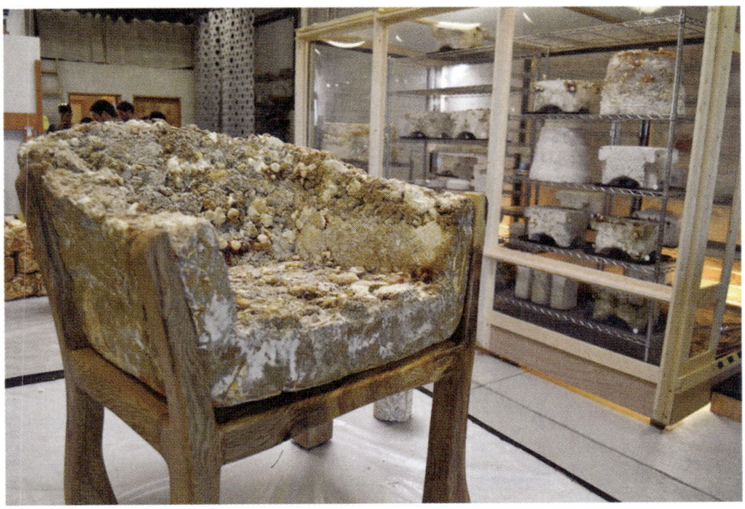

response of the material, rather than as an appeal to programmatic thought. While this approach shares issues in common with digital devices, we argue that this responsiveness opens the door to a wider potential, where proximity and even communion of inter-agents — be they human or material — will take forms of great intensity. Of course, the particularity of these responsive materials may be combined with the play of the device at a more conventional digital interactivity, and the combinations for exploration are fully open.

Metamorphosis, matter made flesh

A material gifted with intelligence, with a sensitivity to the world that surrounds it, extremely gentle perhaps, or with faculties that amaze us. Able to move, to evolve, to transform. To take shape, like the living, which served as its model. Our deep desire with this material is to cross that thin gap which still separates it from the living. A passage between imagination and reality, one that artists love to cross.

The diversity of forms of metamorphosis in literature and the arts are endless, beginning with the founding myth of the creation of the world and men in the Bible: for "the Lord God moulded man from the clay of the ground." (Gen., I, 7) and "from the rib that He had drawn from the man, the Lord God fashioned a woman." (Genesis, II, 21). This is the subject of Ovid's epic poems in *Metamorphosis*: metamorphoses crossing the gods and the mortals of mythology. They are found in the traits of the beetle in Kafka's *Metamorphosis* or Ionesco's *Rhinoceros*, *The Portrait of Dorian Gray* by Oscar Wilde, or *The Strange Dr. Jekyll and Mr. Hyde* by Robert Louis Stevenson. Children's tales and literature about with such stories: *The Adventures of Pinocchio* by Carlo Collodi, *Alice in Wonderland* and a score of others.

Metamorphoses are of all kinds: alchemical, with the transmutation of lead to gold, organic, with the concept of the metamorphosis of the ideal form in Goethe's theory of plants, or mythological and religious. Dust is modelled into mankind, gods turn into men, men into animals, statues come to life and the living petrifies. And of course Jesus the Son of God descends to earth to share in the human condition. He left the world of spirits and purity, to share our situation as men of flesh. Renouncing the fantastic for a foothold in the reality, the concrete, the material of our world. Twice.

The responsive material comes alive, then. It is no longer modelling clay or carved stone that waits for the good will of Gods who are alone capable of infusing it with anima. It is endowed with autonomy, in its way, leaving its status as object to become a fully-fledged subject. About its actions: it moves, but not randomly. It follows its own mobility in a sensible relation to the world that receives it. Its intelligence is structural, stemming from the depths of its intimate constituents, building on and exploring the space in which it opens or closes.

This is not a machine. (More simple or more advanced than that?) What are its challenges, its objectives? Faced with the void onto this point, without the non-substance that makes life (cruel disillusionment), art likes to complement, to extend the illusion. The artifice. First the artist and then the audience after him, project on animated material their own mobility in order to better dream and believe. Alterity always needs to rebuild itself, and it is so easy to put on that dress of the "other" in this medium that lives without a soul. Always re-staging this performance, this metamorphosis, transitioning in continuity, is this change in form or nature, of one state of being to another, from fertilized egg to quickening, from the foetus in the mother's womb to the baby, from primitive differentiation to otherness, from adolescent to adult, from individual to society.

It is not form that is modelled here but the outline of the other, that we agree to see. An other, yes, but another us, an alter ego in short, who necessarily thinks like us (how could it be otherwise, given we are the original). An equal that we raise up despite its material status, to bring it to eye level and see ourselves in it, like Narcissus sees his image on the surface of the water. We follow its movements and gestures, the gestures of the material, the mental gestures, its doings, its life. Here we are in the situation of one who believes, faith being the way to escape the idea of death.

The digital introduced a new modality: behavioural algorithms that allow sensitive objects — rendered on screen or presented in physical space, to be aware of their environment and behave in an appropriate and coherent manner with this world. An art of artificial life. Indeed, the digital can be fantasized as the tool to achieve a universal description of the universe and all life of which it takes part. A tremendous program! A tool in any case that invented a site of "co-active life companion, even acting by delegation.[117]" The behaviour of this digital media is defined so its articulation will be relevant in bringing to each what it needs. The proposed form of life is conceived here on a perpendicular grade to that of responsive materials, that of

[117] AZIOSMANOFF Florent, *Living Art, L'art numérique*, Paris : CNRS Éditions, 2010, p. 21.

an artificial nervous and neural system, with a reduced range of input and output parameters. The computer has been designed in line with the progress of machines, the maccchinali elementi of Leonardo da Vinci, of celestial clocks, astrolabes and other automata, to perform a task independently to assist or replace the human. Tasks are listed and then processed according to a program that governs this great *Whole*, a *Grand Computer,* keeping control.

The materiality in its complexity can be introduced to form new couples of human-device with modular degrees of reliability. Responsiveness would incorporate in a complex manner both digital and analogue levels. The operation of the device, in which the human is an involved party, would be the result of a kind of homeostasis through continuous iterative corrections. It is possible therefore to imagine a "computer" analogue, produced by emerging materials, creating systems of gifted behaviour. Similarly some engineers have sought to translate logical operations via analogue systems such as digital microfluidic systems based on air bubbles in liquid (flip-flops, or ring oscillators)[118], these systems can obey their own logic without going through a program or through centralized synchronization[119].

No hierarchization of information therefore, but rather unstable dynamic equilbria, played globally, from a set of local specific responses. Not machines but materials capable of approaching the boundaries of the body and its emotions... In fact, the responsive

[118] See *Bubble Logic* de PRAKASH Manu, available at:
http://fab.cba.mit.edu/content/projects/bubble-logic/

[119] LABRUNE Jean-Baptiste, "Machines analogiques vs Machines numériques: Deux visions de l'altérité", Planète Laboratoire, Gaité Lyrique [online], 17 octobre 2011 available at:
http://www.gaite-lyrique.net/gaitelive/machines-analogiques-vs-machines-numeriques-deux-visions-de-l-alterite.

8. Animal matter

a- A robot able to walk on vertical walls, like a geko. Its feet covered with a material that adheres because of its microstructure, identical to that of the animal's feet.

b- Structure in strips of the geko's feet, allowing it to take advantage of Van der Waals forces to strongly adhese to walls.

c- Microstructure of a butterfly wing. Its characteristic iridescent blue color is due to the particular structure of its scales and not the presence of pigments.

d- *Tape Vienna*. Collective Numen / For use. 2010. The monumental sculpture of transparent tape is created in situ, like the cocoons of caterpillars or spiders (about a 35km strip). It is strong enough to support visitors, who can crawl inside.

e- A gold colored dress designed with silk thread produced by more than one million female spiders of the species "golden orb," by Simon Peers and Nicholas Godley. Spider silk is a high-performance material that inspire the researchers. Photo Adrian Dennis / AFP

f- *Labyrinth of Memory*. Chiharu Shiota. 2011. An installation made of sixteen white dresses trapped in a black acrylic spider web. A labyrinth where visitors immerse themselves.

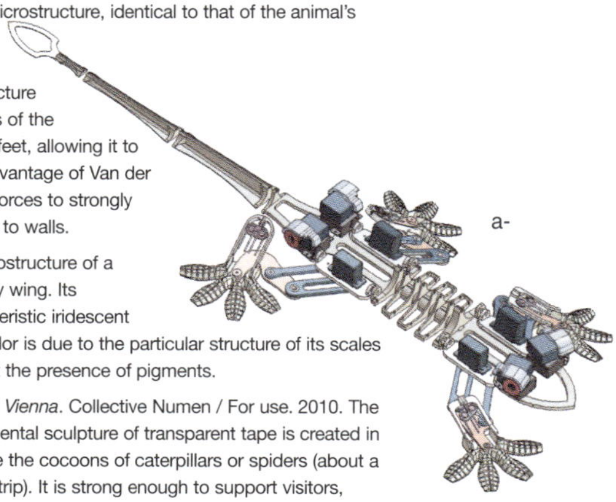

a-

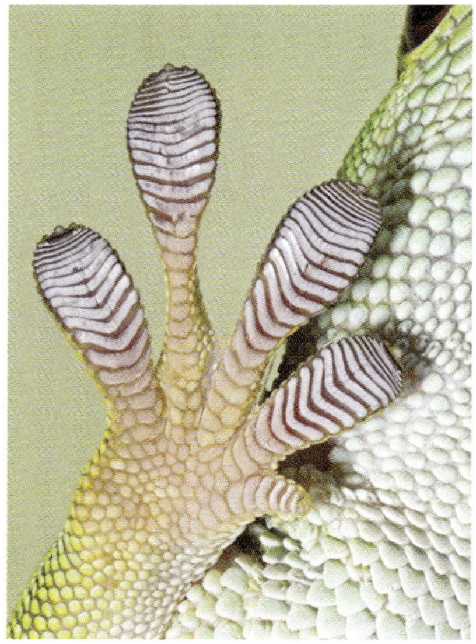

b-

c-

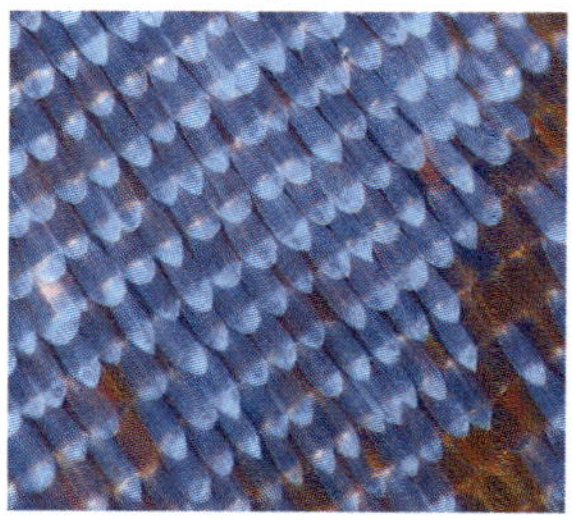

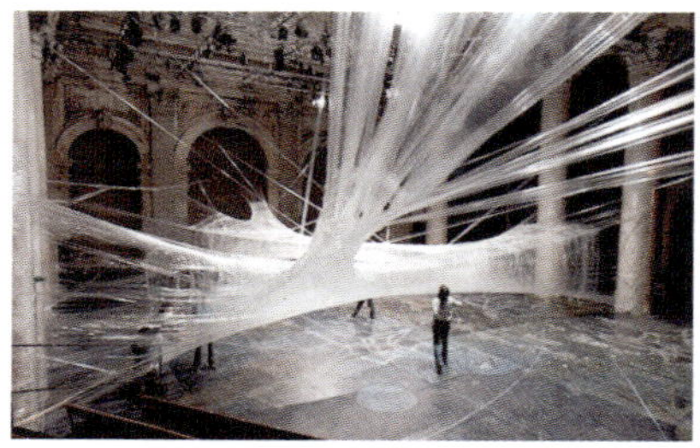

d-

e-

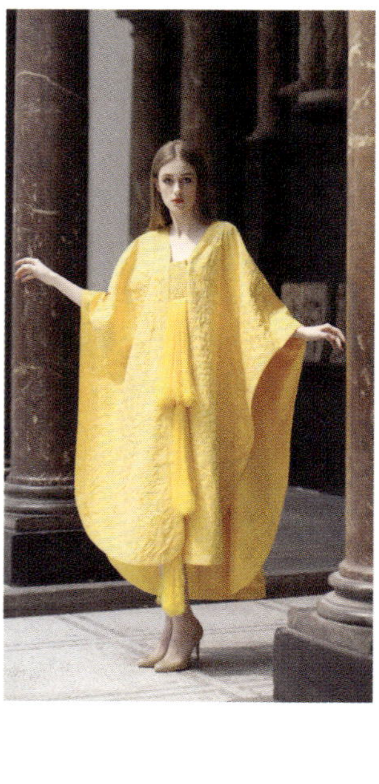

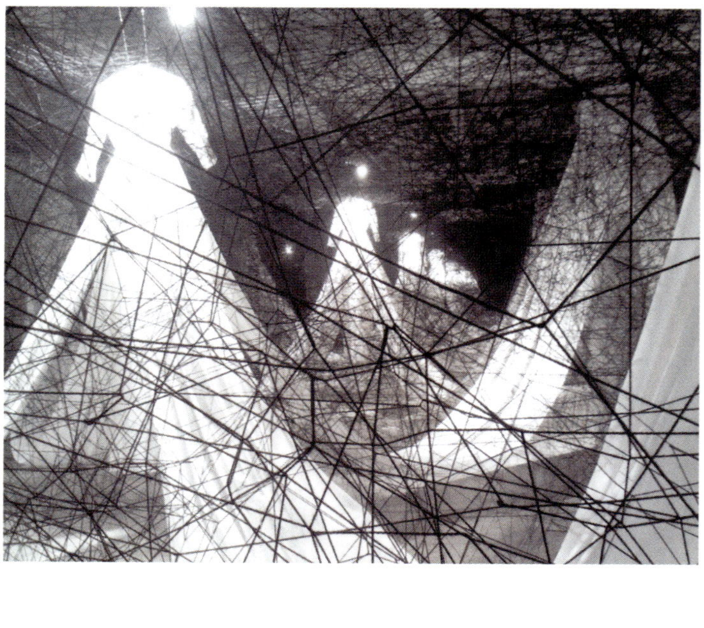

f-

materials move the relation between public and work to the side of touch—the glorification of which cannot be better evoked than through the myth of Pygmalion[120]. The Greek sculptor, impervious to the charms of women, falls in love with an ivory portrait he carved himself. Venus, who hears his heart's wish, rewards him by giving life to his creation. It is through the caresses of a man in love that ivory becomes flesh. "With a trembling hand he questions the heart: the ivory softens, moved, releasing its initial hardness. [...] She exists.[121]" Alterity is constituted in the myth when the material little by little comes alive under the force of desire. Acting not as alter ego but as love object. At least in appearance. Because finally this beautiful woman has been born from the hands of a sculptor who resolutely turns away from the female. She is not the other, she embodies, on the contrary, the refusal of women. She is him, a projection of self in a body to be loved. The desire is there, however, giving power to this metamorphosis.

The artist, sculptor in the legend but contemporary artist today, gives life to matter in a relationship that cedes to desire. But this is art, not reality: feelings are experienced, expressed perhaps, *act*. Real. But the object of trouble is only an ivory sculpture. As in Greek tragedy, which, to achieve catharsis, reproduces the relationships of love or hate, of shame and resentment into a model network of what constitutes our relationship to others, the material can take forms specific to us so to move or disturb us. And so to transform us.

[120] BERGER Philippe, "Le mythe de Pygmalion et le dieu Pygmée", In *Comptes rendus des séances de l'Académie des Inscriptions et Belles-Lettres*, 24e année, 1880, n° 1, p. 60-68.

[121] OVIDE, *Métamorphoses*, [edited, annoted and introduced by Frédéric Le Blay], Paris, Larousse, 1999, livre X, p. 243-297.

Because disorder is the manifestation of a metamorphosis, revealing a deep self inhabited by forces previously unknown to us and that work without our knowledge[122]. It may be purposefully convened, to lead us to another state, more fragile but more mobile. A place where everything can be redesigned, where the cards may be reshuffled, where order no longer exists, and borders have been blurred: an unstable state but related energy is of great strength. Disorder is also ambiguous — cloudy water is not clear. That which attacks us may be that of desire like acts as fury. A beyond consciousness in any case with which artists love to work as an experience for the public or for their own creativity.

Responsive materials have the characteristic of being both material and driving force. The catalyst of this material-machine IS the material. Mobilizing. The transformed material has this particular power to activate, while also being a stakeholder in the real world. It can tip the balance of inhibition to the side of desire or the project. Because human thought advances by changing point of view, as Alain Berthoz explains. In the beginning is the act: "The act is not the movement, the act is the intention to interact with the world and with oneself as part of the world.[123]" Motivation comes from the feeling the power to magically change the world, seeing it as we think it could be.

The key factor in animating these projects of action is desire. In his later writings, Merleau-Ponty posits desire as the exclusive principle hosting the body image: "The perceiving body," says

[122] As the famous verse from Phèdre's confession well well shows us : "Je le vis, je rougis, je pâlis à sa vue/Un trouble s'éleva de mon âme éperdue/Mes yeux ne voyaient plus, je ne pouvais parler". "I look'd, alternately turn'd pale and blush'd/To see him, and my soul grew all distraught/A mist obscured my vision, and my voice/Falter'd" (1914 translation, Harvard University Press) RACINE Jean, Phèdre, 1677, acte I, scène 3, v. 273 à 276.

[123] BERTHOZ Alain, *La décision*, Paris : Odile Jacob, 2003, p. 9.

Merleau-Ponty, “is not only *also* a body that desires and the theory of the libidinal body is not only an *extension* of aesthetics: perception implicates desire, it is already *a mode of desire*[124].” Animating the systems which constitute the work thus drive us in return: updating our own power, through an experience that we conduct. Desire calls to a state as it *is* but also as it *could be*. To take, caress, complete, synchronize the material, is the order of eroticism.

The material, however keeps this magnitude of possibility which can invite interactivity much more attractively than the digital, to become deep anxiety or to provoke disgust. It can be grey matter, soft matter, but also excremental matter. The *Gloeiobabel Nudeliome Landmoonaia* by Ernesto Neto (2000) provides a convincing idea of the attractive force generated by the material, with bodies that sink voluptuously in a sort of marshmallow scaled to the size of the room, overhung by amorphous forms, kinds of catenary arches in organic fashion. Conversely, other artists toy with the abject, showing what we wish to ignore, bodily fluids and various excretions. *MI(pi) bar* plays with the border between the good and the abject, with a bar project in which urine would undergo treatment to be reused in the form of tea. Claiming, through the concept, to accomplish an “infinite process of absorption, digestion, excretion and re-absorption[125].”

Intersensory intersubjectivities

Touch and haptic vision could restrict us to a relationship of proximity with the material, the surface, without depth of field. That this isn’t the case is because of a certain proximity, a certain *empathy*

[124] DE SAINT AUBERT Emmanuel, “Espace et schéma corporel dans la philosophie de la chair de Merleau-Ponty”, in BERTHOZ Alain, ANDRIEU Bernard, Le corps en acte: Centenaire Maurice Merleau-Ponty, Nancy: Presses Universitaires de Nancy, 2011, p.143

[125] R&Sie(n) ARCHITECTS. Paris, 2007. Available at : http://www.new-territories.com/mitpibar.htm.

with the material was made possible which allows us to take the measure of what separates us. Merleau-Ponty speaks of this when he claims this "sort of ubiquity[126]" necessary to multiply our points of view. The philosopher "feels watched by things[127]." That is, inside each object is a regard, highlighting the critically intersubjective character of its perception of the world. The body schema may also extend into the world and attach itself to objects that become quasi-organs, thus contributing to a veritable "extension of existence." More than this: for Alain Berthoz, things are interpreted by our perceptive thought as being "inscribed in the world of possible actions, from which they take their power to act[128]." He recalls the phrase of Merleau-Ponty, distinguishing two states of being in relation to space: "Either I do not think, I live among things and I vaguely consider space... Or I think, I take space at its source, I now think about the relations which are within that word and I realize that as they live only through a subject who describes them and who bears them, I pass from the spatialized space to spatializing space[129]. "

This back and forth between different points of view is justified according Berthoz by neuroscience research which shows the existence of two methods of observing space: allocentric or egocentric[130]. It is possible with strategic mind control to move at any moment from one space to another more suitable, or even to summon them both simultaneously. Or to have an out of body experience that allows for the empathy needed to put oneself in the place of the object. This appropriation of the body by the mind, which we can project to

[126] MERLEAU PONTY Maurice, *Phénoménologie de la perception*. Paris: Gallimard, 1945, p.295.
[127] MERLEAU PONTY Maurice, *Le visible et l'invisible,* Paris: Gallimard, 1964, p.183.
[128] BERTHOZ Alain, ANDRIEU Bernard, Le corps en acte: Centenaire Maurice Merleau-Ponty, Nancy: Presses Universitaires de Nancy, 2011, p.16
[129] MERLEAU PONTY Maurice, *op. cit.*, p.281
[130] An egocentric reference system is that which is centered on itself and an allocentric system occurs when the representation of space is independent from the position of the observer.

see ourselves better, as being integrated in the world, is something very powerful. And archaic. Very powerful, because very archaic. This vision of perception revisits certain forms of animism, for example, as defined by contemporary anthropologist Philippe Descola, that is to say, the perception of a common identity of interiorities between subject and object (living or not) but a distinct identity between their physicalities.

The works of responsive materials thus build a relational fabric, complex but strong, between us and the device, into which our imagination likes to blend, to continuously extend our body schema by the representations that we privately make of this material.

PART III — SOCIAL MATERIAL

1. META-MORPHISM

The notion of order and the structure of thought may also be considered at the scale of society, as the configuration from which we all think. Michel Foucault characterizes this arrangement by distinguishing several levels: the empirical order composed by fundamental cultural codes; the order established by reflexive knowledge (scientific theories and philosophic interpretations); and, finally, a less explicit order, more confused and obscure but more archaic and fundamental. This is the order that seeks to translate across a culture, a space from which knowledge of a period is constituted, which is its condition of possibility: the *episteme*[131]. This is obviously not an objective or fundamental order but a configuration that establishes a substrate from which we think and which can undergo transformations and discontinuities through the course of history. It is important to emphasize that this is an implicit structure, of which we are unaware. Art allows us to see and feel such intrinsic perceptual structures. For, according to Umberto Eco, it is possible to find in the underlying structure of artworks an epistemological metaphor: "In each age, the structure of various forms of art show —

[131] FOUCAULT Michel, *op. cit.*, p.13

a- 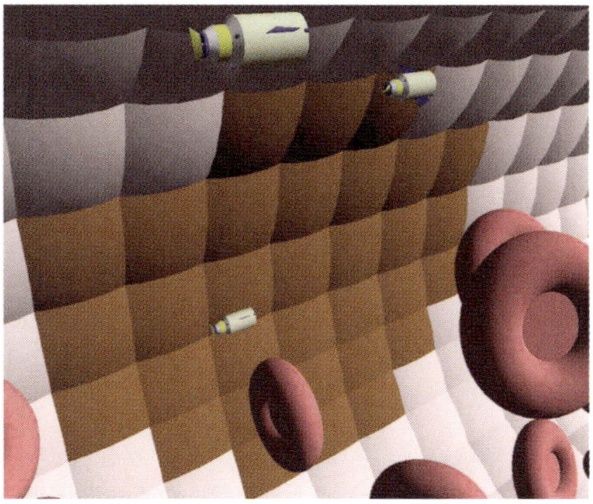

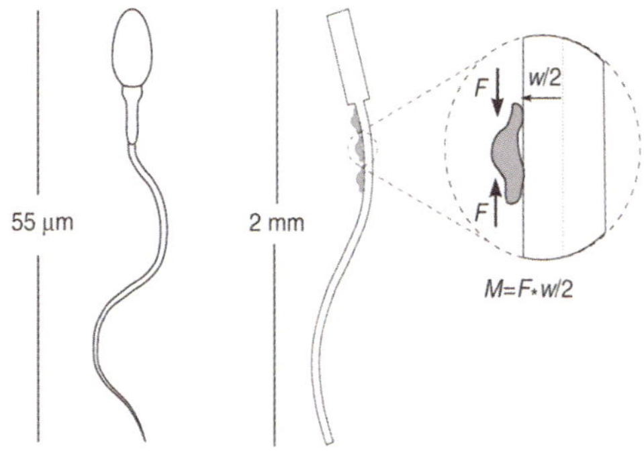

b-

c- 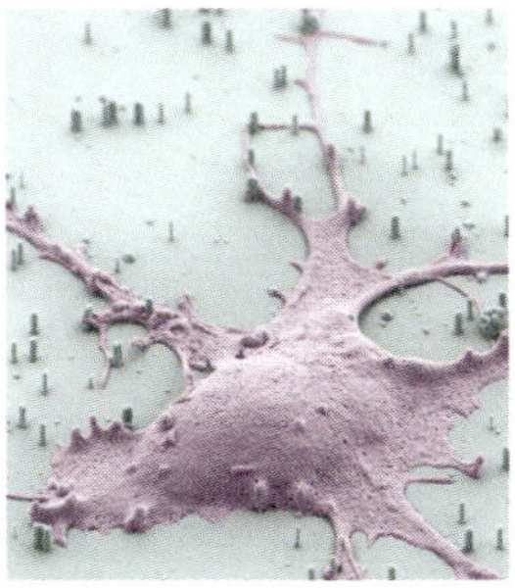

9. Invasive invasion

a- Nanorobots as they could be made to fight against infectious epidemics by introducing them into blood circuits. In Cavalcanti, A.; Shirinzadeh, B.; Zhang, M .; Kretly, L. C. NanoRobot Hardware Architecture for Medical Defense. Sensors 2008, 8, 2932-2958.

b- Morphology of a spermbot (sperm robot). The system can navigate even in viscous fluids in the human body. University of Illinois, Taher Saif and students.

c- Greatly enlarged image of a rat neuron on a bed of nanoscopic peaks introduced to deliver molecules into cytoplasm. The goal of these scientists is to emulate and better understand the brain in order to create more powerful computers. Pr. H. Park, Harvard University.

d- 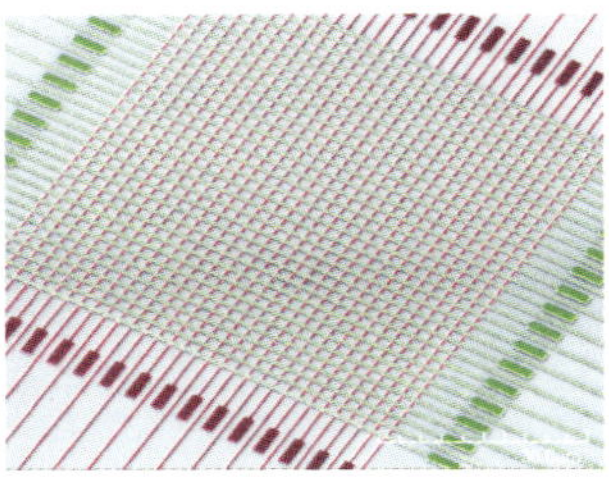

e- 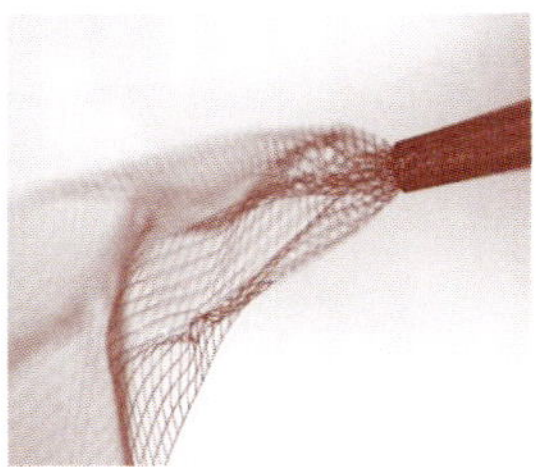

d- Researchers have developed material for making a "memistor" (memory-resistor). Just as the synapses of real neural networks, when a signal passes along a certain path, it is then easier to pass again on the same path. UC Santa Barbara and Stony Brook University.

e- Charles Lieber of Harvard University has developed a net which can be injected in the brain with a simple syringe to record brain activity or even stimulate certain cells without having to introduce electrodes.

f- 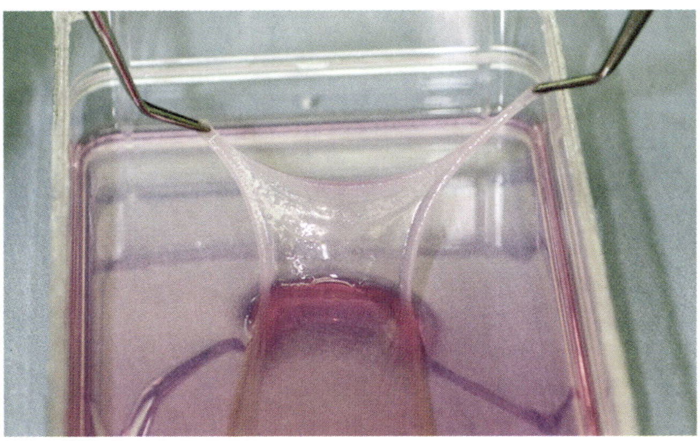

f- C. Baldeschi and M. Peschanski from Inserm and Genethon have transformed human embryonic stem cells onto epidermis for the treatment of burn victims.

in the broadest sense, through similarity, metaphor, the resolution of the concept of the figural — how science or, at least, contemporary culture sees reality[132]."

In his book, *The Scientific Imagination*[133], researcher Gerald Holton, who holds both a chair in physics and a chair in the history of physics at Harvard University, addresses how imagination is put into action by scientists when they make discoveries. He describes fundamental moments of the scientific process using concrete cases from the history of science in order to identify their psychological and social determinations.

Holton initially specifies that the intellectual investment of hard scientists is particularly intense because of the charismatic power of the activity[134]. He examines how the imagination may be governed by adhesion, conscious or not, to a thêma or to thêmata: first convictions rooted in the depths of being. One of the essential functions of a thêma is to render the world intelligible, "in a way that the imperatives of logic alone would not afford[135]." For Holton, famous scientific discoveries obey such thêmata and the controversies they generate crystallize around two or three thêma : the number is ultimately quite small[136]. It consists most notably as antagonisms: "atomism/continuity, simplicity/complexity, analysis/synthesis, and invariance/evolution/catastrophism[137]". All postulates for understanding the training methods of schools, of doctrines, of debate.

[132] ECO Umberto, op. cit., p. 28.
[133] HOLTON Gerald, *L'imagination scientifique*, Paris: Gallimard. 1981. (Bibliothèque des sciences humaines)
[134] *Ibid.*, p. 12.
[135] HOLTON Gerald, op. cit.
[136] *Ibid.*, p. 30. The number of themes is also restricted in physics: according to Holton, their recension allows establishment of about a hundred.
[137] *Ibid.*,p. 9.

But these thêmata, as sovereign and inclusive they are, do not always do a disservice to scientists because they may be "one of the primary sources of energy for sudden innovation[138]." Holton's illustration of his thesis includes a text written in 1967 by scientist Steven Weinberg on the constitutive principle of the nature of matter[139]: researchers striving to find a common basis for the four different forces which account for all physical phenomena – the gravitational force, the electromagnetic force, the strong nuclear force, the weak interactive. An ambition motivated by an investment that is particularly powerful among many theoreticians.

According to Goodman, artworks should help increase and expand the potential of our perceptual and cognitive abilities, enriching the images and predicates we use to see the world[140]. Allegories and metaphors are the pictures and stories that can elicit new concepts in the *episteme*, thanks to their power of transference[141]and transduction[142]. Our perceptions of the world are limited by the opportunities we grant to our own symbolic production. The work of artists with emerging materials will not necessarily facilitate comprehension or rationalization, but rather generate new points of budding, as mental entries to every floor, walkway and panorama. Thus, they may increase the complexity that we are capable of admitting by introducing the unknown. Stretching the episteme, so that we can leave the marked paths. Doubtless, too, we might also rise to

[138] *Ibid.*, p. 11-12.
[139] *Ibid.*, p. 31-42.
[140] GOODMAN Nelson, *Of Mind and Other Matters*, Cambridge : Harvard University Press, 1984 ; *L'Art en théorie et en action*, Paris : Éd. de l'Eclat, 1996.
[141] The capacity to form analogies, that is to say to transfer one situation onto another
[142] Transduction is a kind of reasoning typically used by children from the ages of two to seven, before induction and deduction, notably used for syncretic thought. For Simondon, transduction designates the operation of using form to explain the genesis of the individual based on pre-individual reality. The word is used here in the first sense while taking on the connotations developed by Simondon.

a-

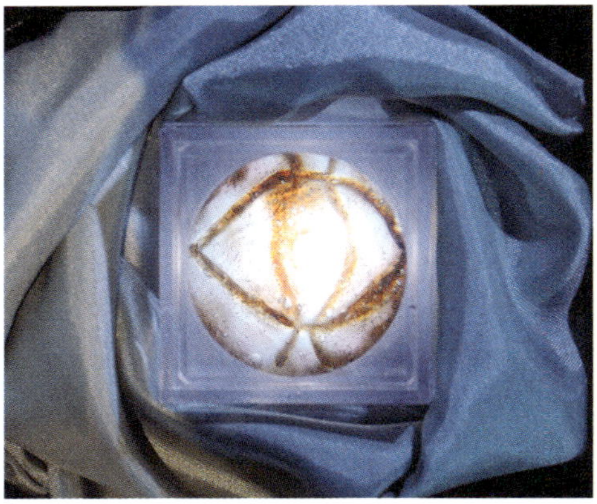

b-

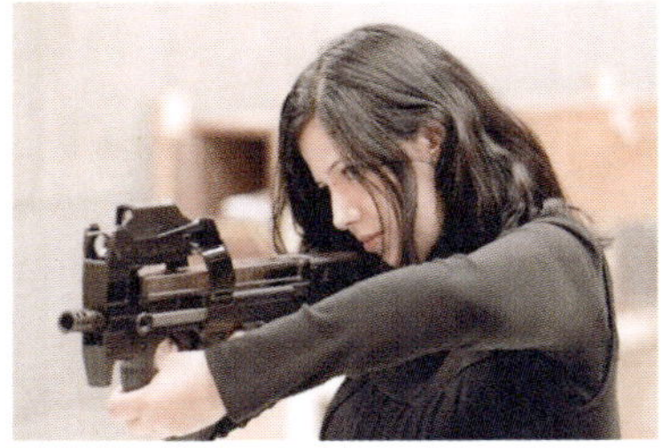

c-

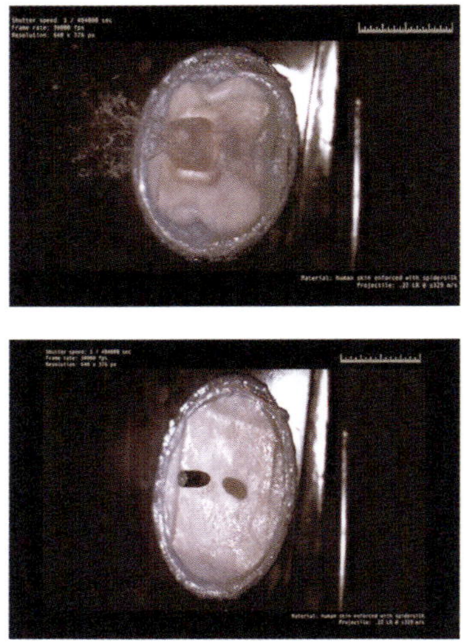

10. Trans-mutation

a- *HymNext Hymen Project,* Installation. Julia Reodica, 2008. Hymens synthesized from regenerated skin are presented as "precious gifts" exchanged between lovers, questioning the traditional value of the hymen now that it is possible to regenerate such skin surfaces.

b- *2.6g 329 m / s*. Jalila Essaidi, 2010. Human Skin hybridized with spider silk, subjected to shooting by a 22 Long Rifle. Could the creation of a hyper-resistant transgenic skin protect us from the impact of bullets or violence more generally?

c- - 2.6g 329 m / s. Detail.

d- Biohacking. Implantation of a Northstar chip carrying LEDs under the skin of the hand, to generate light when a magnet is nearby. Artist Jowan Österlund's offers to those who wish to produce this kind of intervention.

E - *May the Horse Live in Me*. Performance. Art Orienté Objet, 2011. The artist of the collective, Marion Laval-Jeantet, receives an injection of compatibilized horse blood, and then engages in a ritual of communication with the animal. The artist relies on the feeling of strangeness deep within herself, so as not to forget the importance of "animal part" hidden in each of us.

d-

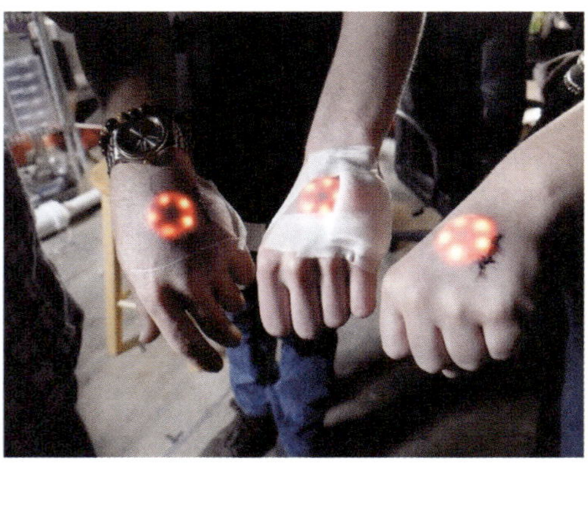

e-

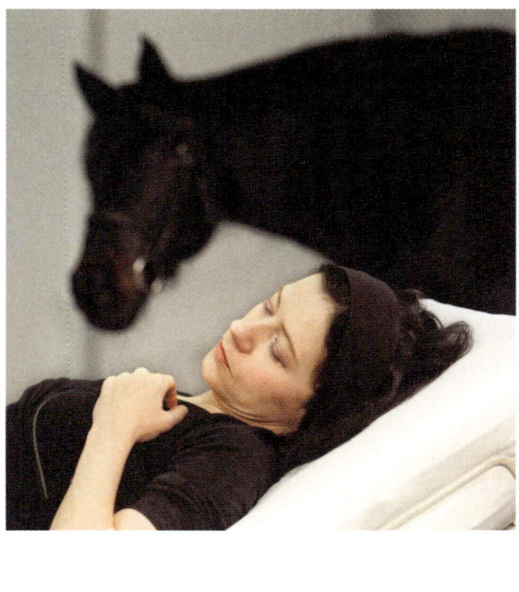

see what's below us, or, contrarily, dig a bit deeper, suspecting our precepts, stalking triviality, connecting concepts.

A work of thought which may be done ahead of logic. The artist in this case can join precursor as defined by Lacan[143]: not the one who goes ahead of categories not yet created but which will come later. But the one who sees, in what is taking shape as thought, conscience, action, technique... the way that they will be perceived a century later. On condition they avoid the risk of contamination or distortion through dangerously clumsy interpretation and don't play on simple *fetish formulas* extracted from scientific documents without understanding.

2. REPRESENTABILITY

To engage several interacting subjectivities is to face this limitation inherent in distinct modes of perception: that of the self, intense and confusing, and that we have of the other, clear but distinct. The outline, the border that separates me from "others" or the "other" does not necessarily stop at the boundaries of our corporeal envelope. The limit of imagination, for example, between ours and that of others is not so obvious... The relationship of the subject to the real and the imaginary and its relationship to the collective and the individual re-emerging continuously. A way of questioning who is the Other in me, or who I am in the other where I look for myself.

[143] LACAN Jacques, *Le Séminaire Livre II, Le moi dans la théorie de Freud et dans la technique de la psychanalyse*, Paris : Le Seuil, 1978, p. 45.

Appropriation of emerging materials by artists may provide some responses to these questions. Because it is a vision of society, past, present, or future, which is expressed throughout the practice of an artist, that questions the relationship between self and other, individual and collective, private and public. But perhaps even more than this – if we can get closer to better see the details. Rather than admit the concept of the individual without putting it to the test, artists often point fingers to where things are less well-defined than they appear: the deceptively simple. More fragile and more complex than one might believe, the interfaces between individual and collective are porous walls. For Mondrian, in his time, the work of individual art was losing its singularity in favour of the creation of a new space, including the architecture of the city, of houses, their interiors, their furniture and daily utensils. With the hope that the *transindividual* might take precedence over the individuality of the one. Later, Beuys, marked by Nazism, sought instead to trigger the rejection of any unifying ideology, like communism or capitalism, inviting individuals find a creative and singular social mode for themselves. His works, both in their individual relationship with the public and through their supposed roles as social sculptures, set in motion a fluid transindividual zone (where the individual is prolonged and surpassed) with the goal of agitating forces that will enable process of transformation. Their lack of style was championed by the artist as not imposing his ideas on the public but allowing them off the bench to generate their own.

Individuation

We are not accustomed to asking ourselves questions about our individuality: we are each *one*, inseparable from our body and identified by our relationship to others. To take things at their naissance, self-awareness is not innate: the process of individuation occurs from a very young age, creating an ordering that determines the realization of a complete and completed individual form. But also leaving a trace, so deeply archaic and buried, of when we were the world and the other at once. Individuation thus requires an inner transformation, taking into account the contradictory and conflicting elements that will define the psychic totality. To individuate is to take the path of one's own fulfilment.

The encounter with art, or rather with *one* particular work, may participate in this work, or at least be inscribed in it, to find its place there, to reveal it or become its tool. For it is always an individual meeting, "Art happens one on one[144]." The work deeply touches our being and this is why it grants access to a kind of freedom, outside the group, outside society. The work that speaks to us above all is that of an artist in all her singularity, that is, someone who has stepped to the side of society to escape its globalizing effect. Emerging materials allow the establishment of an intense relationship with the public to levels that can be very archaic. Guided by her own tensions, the artist can find in the emerging material the means of expression that will enable our process of *introjection*[145], that is to say, to separate between what we find unpleasant and what is good for us. Such as being content, focused, is what lallows us be lulled literally or figuratively by skin-work or enveloping work. But they may also invite us to go into poetry, to follow narration, to access symbolization: in short, to participate in the development of a discourse about their subject. And thus, to separate us. A renewed commitment to balance

[144] SOLLERS Philippe, *L'art réfractaire au social* [personal website], [accessed on february 22 2016]. available at: http://www.philippesollers.net/textes.html.

[145] Introjection is a psychoanalytic term, referring to the process by which a person is able to unconsciously internally integrate those elements that are beneficial or seductive.

confusion and lack of distinction on one side and separation and distancing on the other. For Serge Tisseron, the aesthetic pleasure that we find will be "to the extent of this back and forth, this series of confusions and diffusions[146]."

But once individuated, what role can be maintained by the transindividual? According to Gilbert Simondon, no individual — regardless of age — can exist without an *environment* that results simultaneously in this operation of individuation and which is its complement. That is to say, psychosocial individuation is both psychic *and* collective. This *collective* is neither a gathering of entirely separate individual psyches nor a pure social state without individuality. There are many customized and unique psyches in each group and individuation in a *collective unit*[147]. Simondon addresses this notion of environment by establishing a direct analogy with the material[148]. It evokes an effect that can be found in condensed matter: a closed feedback loop between environment and individual. Finding the movement consistent with its nature may conflict with an opposing force, the alienation called forth by a medium which tends towards uniformity and the passive anaesthesia of the masses. The association *Ars Industrialis* extends somewhat Simondon's notion of individuation by introducing technology as the agent allowing for psycho-socio-technical transindividuation. The social is produced by participation in environments where meanings are formed from the beings that make them as well as from the techniques and technologies that are

[146] MISSONIER Sylvain, "Entretien avec Serge Tisseron", *Le Carnet Psy* n°73, p. 26-37. Available at : http://www.carnetpsy.com/.

[147] See the definition of "transindividuation" proposed by Bernard STiEGLER on Ars Industrialis website, available at :
http://arsindustrialis.org/vocabulaire-ars-industrialis/transindividuation#sdfootnote2anc.

[148] Metastability is a physico-chemical system that is maintained in a sufficiently stable state of equilibrium, but that does not have a more stable state of thermodynamics. For example, supercooled water is in a metastable state.

developed there. The techniques and technologies that permit transindividuation are described by Bernard Steigler as *pharmaka*, that is to say, they can be poisons or antidotes. Individuation would be a form of freedom defined partially in relation to the social body and partially in its opposition. A manoeuvre that is not the easiest, because, as Philippe Sollers says: "This is of course what society is unable to understand because society is primarily occupied in organizing the masses, the people.[149]"

Philosopher Marie-José Mondzain speaks of the need for collective interrogation "about the political destiny of our emotions."[150] Sociologist and philosopher Gabriel Tardes has already attributed to artistic production such a social role, recognizing the propriety of giving "an individuated form to something that would otherwise remain in the order of the pre-individual"[151]. Sensations properly put have a dynamic of assimilation by society through imitation occurring more slowly than ideas or wishes, done at the risk of tipping modern democratic societies towards the lack of their inclusion, or worse, manipulation. Works of art have this property of being able to touch the areas of human experience whose full *introjection* is difficult, if not impossible, due to the too-strong emotional charge that accompanies them. This is the case especially with trauma, individual or collective, such as that caused by the destructuring violence of the attack. They act via two main modes of operations: envelope operations and processing operations, while mobilizing issues that go beyond the content specific to each.

[149] SOLLERS Philippe, *op. cit*.
[150] MONDZAIN Marie-José, *L'image peut-elle tuer?* Paris: Bayard, 2002, p.87
[151] ANTOINE Jean-Philippe, "Tarde, commun sensationnel" *in* Multitudes. T. 4, 2001, n° 7, p.203-211 [1]

For Professor Ronald Hanson, director of the team that achieved the first successful teleport in the world, "If you assume that we are nothing but a set of atoms connected to each other in a particular way, then in principle it should be possible to teleport ourselves from one place to another." Published at LePoint.fr

The development of work on the individual and the collective is essential as part of an ethical and social interrogation, not only *with* the material but also *on* the material. High-tech materials provide access to new functionalities in areas as yet obscure. The field of emerging materials is definitely at the threshold of an exponential evolution. Already fed by important theoretical ambitions, as the principles of molecular engineering now made possible by the latest technological innovations are opening the field to systems which become more and more strange and disturbing: those of the "uncanny".

This new technological material revolution holds great potential for applications in virtually all areas: health, the environment, energy production, agribusiness, design, architecture, etc. Work on the front line between the living and the artificial is underway in many laboratories. Significant changes in economic and social fields will soon be felt, perceived by some as the source of benefits and future enrichment, by others as the bearer of negative effects, and even severe and uncontrollable societal risks. What is certain is that emerging materials will transform our grip on the real world, either to take greater control, as with modified DNA or synthetic blood, or contrarily, for cutting oneself off from it, as with Invisibility cloaks. "The history of the body can be seen as a history of machines,[152]" says Eulalia Perez-Sedeño, who reminds us that the desire to prolong life expectancy is proof that we are moving towards transformation and mastery of nature. It is nature itself which, once transformed artificially, could become a technical object. The technology of emerging materials is definitely in the order of the *pharmakon*, literally and figuratively: perfecting the body by way of cure or poison. Transforming our relationship with nature, for better and for worse.

Apprehension of emerging matter according to the classic couple of form/matter does not, as we have seen, account for its modes of genesis and its operations. Although media coverage of exceptional applications is significant, it focuses on some very specific discoveries because they are easier to understand and more enticing, forsaking all others considered less "sexy" in media-speak. Even more, this transforms the reality of presenting proposals so that each is crazier than the one before, like the announcement of possible human teleportation, discussed in the summer of 2014[153], leaving our capacity for judgment at the door. A teleportation which, when it comes to the experiments performed in laboratories, remains light years away from the representations imagined by science fiction. At the point things stand, teleportation simply acts to carry the *quantum state* of a photon over long distance optical fibre[154]. Media descriptions of the properties of these materials, often very superficial, don't allow actual

[152] PEREZ-SEDEÑO Eulalia, "Ubris technoscientifique et corporéité: l'enjeu des nanotechnologies", papers from the seminar, *Perfection et perfectionnements du corps* [on line], 5-6 January 2009 [accessed on 8 October 2014]. Available at : http://s2hep.univ-lyon1.fr/Pefect_corps/participants.htm.

[153] The term teleportation inspired journalists: Professor Hanson's team has demonstrated that it is possible to teleport information encoded in particles between two points three metres apart. Science fiction then took over the reporting.

[154] This is an effect of quantum entanglement, a phenomenon arranged by subatomic physics. To learn more, see, Nicolas Gisin "L'impensable Hasard, non-localité, téléportation et autres merveilles quantiques", publisher: Odile Jacob.

understanding of the processes involved, which remain un – or poorly – symbolized. Here, art can perform its historic function of producing a proper imaginary for this new material. To generate different comments and perspectives, to build cognitive frameworks athus far unknown and to imagine links between the public and the scientific community. To produce meta-narratives on the new material, to better understand it and to initiate discussions on ethics of its implementation.

Exit the master-slave relationship

Although we agree with the position taken by the great epistemologist Bachelard in his statement on the individualizing power of material, we do not adhere to his notion of a uniform materiality across all size scales. It is true that this often applies to the material components of our world: water, earth and fire. It seems to us that it can be just as powerful to finally abandon this vision of matter as amorphous and easily adaptable, as potentially separable into many identical parts, making each a whole.

Because material IS *form*, in a way internal to itself. And because of this, its *formatting* is the source of its properties. Cutting or dividing can annihilate everything. Deformation transforms its properties. Therein lies its power. Laboratories and industry thus often focus on formatting and structuring. But if it just boiled down to their chemical formulas, the list of materials would be finalized quickly enough and we would not be discovering more amazing properties for years to come.

Simondon has already proposed leaving the old hylo-morphism of material/form. He associates this with the "master/slave" relationships dominant in antiquity[155]. We have passed into another scheme with dominates our complex and globalized social relations, in whole and in part. To best capture our conception of complex material without reducing it to only the couple of material/form, it is probably necessary to observe its development, the process of its genesis, and its operation.

But as rightly noted by philosopher Sacha Loeve[156] specifically on the topic of nanotechnology, while these subjects "have been intensely mediated, nanotechnology practices are not symbolized and their occultation is the basis of the rhetoric and the communication of nanotechnology.[157]" The researchers themselves have controlled the material's molecules more and more in order to manipulate them individually and thus get closer to the process of genesis, if their work can be realized. But, says Loeve[158], communication about these modes of material individuation is very bad and the nano world fails to translate them to those outside that world. This leaves the process of the individuation of these ultra-sophisticated materials in a grey area, if not in total darkness. The idea that the molecules just enter a nano-machine from one side and exit the other (nano-gear, nano-engine[159], nano-vehicle), continues. This results in discourse which, although having a foothold in the reality of the practice, does not coincide with it[160]. The question of the

[155] DÉOTTE Jean-Louis, "Les Immatériaux de Lyotard (1985) : un programme figural", in *Appareil : Lyotard et la surface d'inscription numérique*. 2012, n° 10, p. 3-4.
[156] LŒVE Sacha, "La zone obscure des nanotechnologies", *Appareil*, 2008, n°2. Available at: http://revues.mshparisnord.org/appareil/index.php?id=635.
[157] *Ibid*.
[158] *Ibid.*
[159] A dynamic representation of a molecule-motor-rotary is available at: http://www.cnrs.fr/cw/dossiers/dosnano/.
[160] LŒVE Sacha, *Ibid*.

representability of high-tech material is paramount if we want to take hold of the debate on their functions, uses, and development.

Raising a popular opposition against GMOs in France has had the effect of mobilizing many initiatives, of pro-GMO media campaigns to strong anti-GMO responses, through citizen debates such as the creation of *citizens' conferences,* the first held in 1998 on the subject of transgenic plants[161]. Thanks to these violent oppositional movements which have overtaken their industrial and political protagonists, research on nanotechnology and its perspectives were accompanied by very large budgets devoted to media and for the organization of scientific and civic watchdogs[162].

But risk management has often been accused of addressing social acceptability rather than providing real awareness and knowledge. Participation of the public has sometimes been bogged down in complicated procedures which eventually lead to soft consensus. The difficulty lies in taking stock of what nano-technologies as such genuinely involve, because of the complexity of nano systems, the devices and procedures for citizen participation generate the construction of standard narratives and standard discourse. Ultimately, the lists of problems are at once urgent (as nanomaterials are on the market) and totally futuristic (as in the concept of transhumanity), largely hijacking political debates about crucial ethical and scientific questions[163].

Sacha Loeve has noted, for example, three great strands that traverse the discourse around nanotechnology.

[161] Organized by the l'Office parlementaire des choix scientifiques et technologiques (OPECST) with 15 citizens.
[162] For example, OMNT is an observatory that provides scientific monitoring of important topics in micro and macro technologies.
[163] LAURENT Brice, Les politiques des nanotechnologies. Paris: Charles Léopold Mayer, 2010, p. 11.

The first idea, relayed by media in support of the nano cause, is that nanotechnologies are completely natural, given that living systems are themselves a structation of material at the nano scale. Nature is thereby interpreted as a vast nanotechnological processes. A totally misleading argument because the problem is that we manufacture highly reactive nanomaterials which do not correspond to forms found in nature...

The second idea is that of the utopia of the augmented human being. This usually takes the form of an "insular individual cut off from the world and others, who we want to make the consumer dream of tomorrow.[164]"

Finally, the third raises the idea of cryptotechnicity as defined by Simondon: the *veil* or *disguise* which must adorn technical objects in order for them to enter the *citadel of culture*. The discourse on nanos often rests on the imaginary of science fiction, which floods this gaping void needing representation. A rhetoric that in itself is not problematic, unless — and this what Loeve tells us — it is exploited for the purposes of biased communication.

Art Experiences and OpenLabs

Various projects have been initiated by laboratories open to the public in the interests of transparency[165]. This is the case of the OpenLab project, which offered five residences to artists for four months in order to raise awareness of the role of chemistry in the nanosciences. Some twenty researchers in chemistry, surface

[164] *Ibid.*, §. 6.
[165] For numerous examples, see: PEYSSON Dominique, *Le nanomonde: une nouvelle "hétérotopie"?*. [plastik] [online],• NUMéRo#03 Nano, February 13 2013. Available at: http://art-science.univ-paris1.fr/plastik/document.php?id=702. iSSN iSSN 2101-0323.

science, materials science, nanoscience, and physics participated in the project. The concepts and techniques of synthesis, such as modelling and all the intermediate stages between chemistry, biology and physics, have been made available and entrusted to contemporary artists. The project, initiated by a chemist, Niki Baccile and a curator and art critic, Margherita Balzerani, resulted in an exhibition, *Invisible et Insaissiable* (Invisible and Unknowable)[166], as part of the International Year of Chemistry. Unfortunately, the funding and provision of laboratories implicitly deflected direct criticism very delicate, but opened the field for imagining new forms rendering the shifted gaze on nano technology. The work previously noted, *The rule of three + 1* by Stefane Perraud with its gold particles in different states of being, partially staged on crucifixes. Other works by Hicham Berrada, Fabien Giraud, Eduardo Kac, Jörg Lenzlinger, Lorenzo Pagliei Michel Paysant, Raphaël Siboni and Gerda Steiner[167] were also shown at Enghien-les-Bains. The forms that were adopted were extremely varied, such as *Aromapoetry*, an olfactory work by Eduardo Kac. The artist used chemistry and materials science in concert to imagine an olfactory poetry from the synthesis and mixing of odoriferous molecules. The resulting poem was captured in a book covered with a silica surface, the porosity nanostructured to remain invisible while containing the molecules of the poem. Raphaël Siboni and Fabien Giraud worked on devices that can more closely manipulate the elementary particles of matter. *La mesure AGLAÉ* (*The*

[166] *Invisible et insaisissable, quand les nanosciences rencontrent l'art contemporain*, 2011 (September 23 to December16), Centre des Arts d'Enghien les Bains. Exhibition organized as part of the international Year of Chemistry, with the support of the Collège de France, the l'Université Pierre et Marie Curie, the CNRS and the Nanosciences ile-de-France.

[167] For a description of works shown, see "Invisible et Insaisissable, quand l'art rencontre les nanosciences", Baccile Niki et Balzerani Margherita. [online], • NUMÉRO #03 Nano, January 15th 2013. Available at: http://art-science.univ-paris1.fr/plastik/document.php?id=657. ISSN ISSN 2101-0323.

measure AGLAÉ) is a video project, in which a camera has been introduced inside a particle accelerator: the Grand Louvre Accelerator of elemental analysis. Exposed to ion beams, the camera sensor was gradually damaged until totally destroyed. The resulting video bears witness to the confrontation between the particle accelerator filming its interior and the camera used for its filming.

META-MATTER

The book *Matière à pensée* (*Food for thought*) by Jean-Pierre Changeux and Alain Connes is written as a dialogue between the neuroscientist and the mathematician. Their exchange brings out one question, to which their responses differ: does mathematics exist *in itself*, outside the brain of the person whose role is restricted to that of discoverer, or is mathematics a pure product of brain activity that could then trace its own structure? In the end, this question led them to ask the ultimate question: is it possible to manufacture an artificial intelligence capable of attaining the complexity of the human, to have mathematical reasoning and, pushing things to the extreme, to be capable of imagination? The eternal question of the duality of matter /mind might finally be solved: a structured and highly sophisticated material, conducting information in the form of electrical particles and chemical molecules according to a dynamic for achieving homeostasis, would be sufficient. The physiological knowledge allowing access through neurophilosophy to the truth of the human without having to bother with metaphysics[168]. This would create a *scientific research of the soul*, conceptualized by *emotional computers*[169] and *spiritual machines*[170]. Artificial intelligence, that of

[168] POIREL Christian, "Le matérialisme neuronal et la question de la transcendance", T.68, in Archives de Philosophie. 2005, p. 77,78.

[169] PICARD Rosalind, Affective Computing, Cambridge : MIT Press, 1997.

[170] KURZWEIL Ray, *The Age of Spiritual Machines. When computers exceed human intelligence*. New York : Viking, 1999.

computer programs, is now outdated, but the behavioural approach and cognitive neuroscience are still tempted by the quest for the Grail, the ultimate matter-mind where brain and material-computer (artificial brain?) could unite to the point of identification. A specific point as well as an infinite space, that of the great connected *Whole* extending like a rhizome without limits and the effects of a border, whose materiality is reduced to that of its lines (in the electrical sense) which allow transfer of intangible data, in essence, being nothing more than the lines (in the mathematical sense this time) of this transportation of immaterial thought.

What a magnificent paradox which calculates asymptotic limits in scientific manner: the human body, made of individuated cells connected by nerve cells to the brain, itself consisting of neurons, would tend by definition towards pure materiality. Living. Nothing then prevents the idea that teleportation could actually work: just reorder the addition of matter as it is ordered here. However, the information flowing along all these cells can completely be considered finally only as a signal. It is therefore possible to imagine functioning in the same way as any other medium capable of carrying information. A way to pure spirit, operating independently from its support[171]. Not without matter, but without corporeality. Or why not a global body, as we speak of global brain? Based on this hypothesis, it would be conceivable to download thought on the neural circuitry of a material-machine of unlimited size, to let it grow according to its own potential. The whole concept of materiality, that of the teleportable body, joining exactly that perfect immateriality, a pure mind that one could upload onto a chosen medium.

[171] After all, isn't this already the case with Stephen Hawking, the great cosmologist, who has lost all his motor capacities? The researcher has motor neurone disease, which has destroyed nearly all his neuromuscular control, including the ability to speak, but which has not prevented him from continuing to produce high level research that contributes to scientific advancement in his field.

The opposition between the materialist and idealist positions courses throughout the history of thought, but the development of new forms of materialism in today's philosophy, such as dialectical materialism, speculative realism[172], etc; are proof that scientific advances upset the known and urge us to rethink (again and again), this rapport between body/mind. Recent developments such as the work of philosopher Quentin Meillassoux pose the question of such a "thought in itself," possibly emancipated from all perception, at the asymptotic limit of disindividuation. So here we are. The scope of a thought without a body is now concieveable. But Meillassoux claims a materialist position — a speculative realism specifically[173] — based on the primacy of matter. Because to speak of the *emergence of life* is to accept the idea of a pre-existant world, a space-time before representation[174]. The laws of physics express themselves, and are even stable, without the necessity appointed by the intelligence of a god or spirit.

However, there have been two or three decades in which it was pretty clear that we were well on our way towards the immaterial. For Lyotard, we were finished in the 80s with having the primacy of materials before us, already there: an era had ended and the future thought would rise to no longer touch the ground and would soar in the post-material. The age was that of the techno-scientific production of our reality, of our language, of our dematerialized art. The exhibition *Immatériaux* curated by Lyotard at the Pompidou in 1985, was conceived as a sensory experience revealing the concept of this post-materiality, linguistic and digital. Metaphysical. A

[172] [173] Speculative thought studies the knowlege of objects which cannot be verified by experience. its practice thus remains solely theoretical, arbitrary, and unverifiable.

[174] "And we discover that this problem cannot in any case be thought through the transcendental, because it concerns a space-time in which transcendent subjects were able to pass from not- having-a-space to having-a-space: therefore, a space-time before the spatial-temporal forms of representation." MEILLASSOUX Quentin, *Après la finitude. Essai sur la nécessité de la contingence*. Paris:Seuil, 2006, p.47.

transition towards the triumph of those products of the mind freed from finitude, Lyotard told us, finally rid the Great Narratives of modernity as fundmamentally based on emancipation.

We could make a de facto clean sweep of this somewhat abrupt description of the material. Instead, we propose to reinvest it in light of the augmented definition by techno-science. Rather than dismiss the materiality, we need to realize how it will mark our future. Materiality is currently the subject of much cutting-edge research that will perhaps make the twenty-first century (post-postmodernism) that of the high-tech material[175]. That is to say, that of a tangible material that have taken note of its passage through the *immaterial*. After experiencing the *immaterial* and its metaphysics, our society imagines the creation of new materials in other ways. There is no longer immateriality on the one hand, materiality on the other, but a continuum, from primordial matter to digital information. From elemental carbon to nanotubes, from organic molecules to metamaterials, DNA to metadata, polymers to cells, cells to brain... Things extend, differentiate, connect. But thus far, they do not announce the limits of materialism.

To build an imaginary through emerging materials can lead to alternatively consider this relationship between matter and spirit. For discussing it in another way, leaving the asymptotic limits of all-body or all-mind to return to its points of origin. The processes acting in these materials may allow us to imagine the genesis of a form of thought as issuing from each intimate part of the body. And from the relationships between them, in all their complexity. Without searching for an analogy, it is possible to understand complex matter per se, as source site for a perception/action sensitive to its environment at all

[175] The twenty-first century could be called the 'Multifunctional Materials Age', from the cover text of the book, *Smart structures theory*, *op. cit*.

size scales, and having a distributed constitutive consciousness. We have seen that in particular each of these primary processes are intimately related to the structure of the material, its macrostructure, its mesostructure, its microstructure - but also to all these structures combined.

Responsive materials reveal a certain tangible *reactivity seeded* in the organization of the material and in its subsequent potential *to act*. A primitive level of response, which may be integrated with an artistic device. No computer system needed, no need to put perceptions of sensors through pseudo-neuronal programming in order to activate in return a process under control. But direct response of the material, according to its own singularity. The artists' stagings give us many images from which we can think about our thinking body. Far from a pure spirit, as well as from an exclusively neural materiality. Here, the material participates in the thinking, the skin of the hand and the hand itself are already thinking, the whole body makes a decision. Intelligence, then, lies not only in software but is hidden equally well in hardware. A thought associated with highly primitive methods, little evolved but extremely robust.

By reversing the process, if our thinking is fundamentally thought-matter, one can imagine that it can easily feed image-materials. Images from complex materials able to support or push our thinking towards a more flexible and broader imagination. For we now have to build a mental space to bring representations of strange and unseen realities. Enlarge the space and make a place in our minds to animate thoughts that put them to work. Metamaterials for example will soon be able to play with our perception by erasing material bodies in our field of vision, objects will be sustained by the way of superconduction, artificial blood will be able to complete (replace?) our blood, stem cells will regrow our defective limbs, chemistry will trade DNA molecules to produce new species... or will one day perhaps deprogram our old age and death like we already know is possible for a certain breed of jellyfish...

The imagination is a political force, as many writers, thinkers, philosophers have shown. Edgar Morin and others have studied how social imaginary could enrich and organize reality; a kind of feedback loop as shown by Simondon and his concept of milieu: the individual coupled with his direct surroundings, which act on him in return. But our thinking has to break with the concept of the individual as the meeting of one form and one material. It is no longer just about taking a shape but sometimes also a transmutation, in order to account for these passages between inert matter and living matter. The myth of metamorphosis, an invariant structure of the human imagination, could be one of its supports. Doesn't Midas turn to gold everything that he touches? Daphne becomes oleander to escape Apollo's amorous pursuit and Pygmalion sees his sculpture spring to life after falling in love...

The staging of emerging materials, exploring all kinds of energy processes – including transformation from one state to another – should be able to carry these metaphorical figures. Expressions materialized of their modes of being and acting can constitute a library of mental images-materials at our disposal. The world of NBIC[176] has already raised some burning questions: how to tackle such problems

[176] Nanotechnologies, biotechnologies, information and cognitive sciences.

if it's impossible to track the causes and issues they present? Accumulate representations, absorb them, assimilate them until they belong to us, until they are a part of us, and then we will be able to think by intuition. Because the problems of the future are complex and multi-agented. Producing the answers to be given to the most serious situations can be done via syncretic thought, more comprehensive and more powerful than pure reason.

The artist of course has to be here. After all, the artificial is her field, she knows it well, handles it, it burns her fingers sometimes (like fireworks), then launches a spectacle in the sky – or lights a tiny candle. Life is a problem she shares with the philosopher, which she plays and replays by changing its staging, by constantly reiterating its repetition. Which is one way to improve, for us and for her, because repetition is a compulsion, as Freud tells us, that allows us to master the trauma, to coax it, for its symboliziation. In art (as they say in the Amazon), reality can marry the artificial and the artist transmute us into transhumans: we can see in the dark, we move according to the Earth's magnetic field, our strength is increased tenfold, our limbs regrow like the tails of lizards, our skin will wrinkle no more, our memory will extend, our neurons will regrow and reconnect. And our thought? Here things get more difficult since by definition we can not yet think what we will think when we think better thoughts[177]. At most we can say we will be different. We will need a corporeal thought to be able to conceptualize the boundary between the inert and the living. The material, emerging, responsive, allows us to access a soma-thinking, which does not necessarily take the road of

[177] In his book *The Blind Spot,* mathematician William Byers demonstrates that we will never have a rigorous theory of consciousness because consciousness is autoreferential. *The Blind Spot*: *Science and the Crisis of Uncertainty*, Princeton:Princeton University Press, 2011. p. 208.

intelligibility. No longer pure rationality, that of the immaterial called forth by Lyotard, but a thought of intuition, on a syncretic mode. It responds with an entirety more profoundly itself. It IS driving force, as a gesture shared between audience and work, between society and material.

The proper images supporting such a dynamic will be even stronger as they hobnob with strange manifestations of the sacred, if they are defined, like Roger Caillois said as "That which gives life and that which rob life"[178]. A common *materia prima*, eminently collective, as Durkheim tells us. Our society has gradually passed the two notions of Progress and Modernity as ideologies to be overcome... and it no longer can be a question of humanism: for the transhuman would flow in and take up all the space. Let us start on other paths of imagination to find the way of our mythology and our metamorphosed myths capable of soothing us.

[178] CAILLOIS Roger, *L'Homme et le sacré*, Folio essais/Gallimard, Paris :1950.